karsten seehafer

dan dragulin

scientific advisor – mircea dragulin

There is no branch of mathematics, however abstract, which may not someday be applied to phenomena of the real world.

Nikolai Lobachevsky

Impressum:

Bibliografische Information der Deutschen Nationalbibliothek
Die Deutsche Nationalbibliothek verzeichnet diese Publikation in der Deutschen Nationalbibliografie; detaillierte bibliografische Daten sind im Internet über https://www.dnb.de abrufbar.

Dan Dragulin, Karsten Seehafer
AI Era of Straightening
Mathematical Foundations and Algorithmic Design

1. Auflage 2026

in der Mediengruppe Westarp
Kirchstr. 5 - 39326 Hohenwarsleben
www.westarp.de, www.westarp-bs.de, www.book-on-demand.de
produkthaftung@westarp.de

ISBN:978-3-96004-184-9

Printed in Germany.

Preamble

The endeavor to understand and formalize the process of *straightening*—the transformation of geometric configurations toward simplified or canonical forms—lies at the crossroads of several profound mathematical disciplines: differential geometry, control theory, optimization, and algorithmic learning. Though the concept might evoke images of physical manipulation or mechanical adjustment, the mathematical abstraction transcends tangible materials, revealing a landscape rich with theoretical intrigue and potential.

From the vantage point of a mathematician, this work is motivated by the desire to explore the foundational principles that govern such transformations in purely geometric and algorithmic terms. The aim is not to solve engineering problems nor to simulate physical realities, but rather to construct a rigorous, conceptual framework that elucidates the essence of straightening as an iterative, self-correcting process governed by deterministic rules.

This inquiry naturally intersects with the modern paradigm of artificial intelligence, particularly in the context of reinforcement learning and adaptive control. However, unlike typical AI approaches that rely heavily on stochastic exploration and empirical optimization, here we seek a direct and interpretable mathematical synthesis. The idea is that AI systems can function as *mathematical agents*, capable of *understanding, interpreting, and adjusting* existing algorithms through explicit, rule-based language informed by geometric insight.

Such a perspective challenges the often opaque "black-box" reputation of AI. Instead, it positions AI as a collaborator deeply embedded in the realm of symbolic mathematics and formal reasoning. This vision suggests that adjustments to algorithms need not be arbitrary or heuristic; rather, they can be the product of precise, mathematically justified commands that an AI can learn to generate and refine.

The mathematical models presented in this book are intentionally abstract. They eschew the complex constraints and irregularities of real-world physics, opting instead for idealized representations: points, curves, manifolds, and surfaces embedded in Euclidean space, manipulated by well-defined iterative schemes. These simplifications are not weaknesses but strengths—they allow us to focus on essential behaviors, convergence properties, and geometric intuitions without distraction.

Readers will find detailed explorations of deterministic self-correcting algorithms, geometric convergence criteria, adaptive learning rates defined through intrinsic quantities, and the interplay between geometric deviation and control policies. The narrative threads a connection between classical mathematical constructs—such as the first and second fundamental forms of surfaces—and contemporary algorithmic paradigms inspired by reinforcement learning, but with a distinctly geometric and deterministic flavor.

It must be acknowledged that this book does not aspire to exhaust the topic. Rather, it embraces mathematical speculation, conjecture, and exploratory models that serve as intellectual provocations. Some formulations may remain incomplete or idealized, intended to spark new questions rather than provide final answers. The goal is to cultivate a fertile ground where mathematicians, computer scientists, and AI researchers can converge, each bringing their unique perspectives to bear on the fascinating challenge of straightening and AI-guided algorithm refinement.

In closing, this work invites the reader to engage with straightening not merely as a computational problem, but as a mathematical odyssey—one that exemplifies the profound capacity of abstract reasoning to illuminate the mechanics of learning, adaptation, and control in both machines and the geometric universe they seek to emulate.

Future Directions and Mathematical Horizons

Looking forward, a rich vein of mathematical inquiry lies in integrating *stability theory* into the framework of straightening algorithms, particularly through the lens of **Lyapunov stability criteria**. By constructing suitable Lyapunov functions—scalar measures that decrease monotonically along trajectories of the system—one could rigorously guarantee the *asymptotic stability* and robustness of the straightening process. This would move beyond heuristic or empirical convergence checks, enabling precise mathematical certification that the system's iterations will reliably approach the desired equilibrium configuration.

Additionally, there is potential in extending the deterministic geometric control schemes toward *hybrid* or *stochastic* models that incorporate uncertainty and noise, reflecting more realistic environments while maintaining analytical tractability. Such models could bridge the gap between pure geometry-driven methods and classical reinforcement learning algorithms, enriching the theoretical toolkit available for adaptive algorithm design.

Another promising direction involves leveraging *differential topology* and *geometric measure theory* to deepen the understanding of how complex shapes and higher-dimensional manifolds evolve under straightening flows. This could uncover new invariants or conserved quantities, suggest novel parameterizations, and inspire algorithms that respect intrinsic manifold structures rather than relying solely on extrinsic embeddings.

Finally, the exploration of AI as a direct programming agent—capable of interpreting geometric data, formulating mathematical feedback, and synthesizing algorithmic adjustments—remains a compelling frontier.

These avenues, among others, underscore the vibrant interplay between pure mathematics and computational intelligence, promising a future where rigorous theory and adaptive algorithms coalesce to produce elegant and effective solutions to complex geometric control problems.

Mircea Dragulin

Table of content

0. Introduction

The study of *straightening* — the systematic transformation of geometric bodies or point configurations toward an ideal, simplified, or canonical shape — stands at the intersection of geometry, control theory, and computational mathematics. Though seemingly straightforward in concept, the mathematical challenges inherent in designing algorithms that can effect such transformations reveal deep and elegant structures, inviting exploration far beyond practical application.

This book presents an abstract, rigorous examination of straightening algorithms, emphasizing their mathematical foundations rather than their physical instantiation. Our goal is to develop a theoretical framework that demonstrates how artificial intelligence (AI) can engage directly with programming languages or existing codebases through *mathematically driven instructions*. In other words, AI need not be confined to opaque "black-box" learning; instead, it can actively *learn, interpret, and adjust* programs via explicit mathematical principles encoded in direct language commands.

By focusing on the essential mathematics of straightening, we sidestep the complexities of real-world constraints and material properties, revealing a distilled picture of algorithmic behavior. This abstraction allows us to highlight how AI systems could potentially perform direct program modifications informed by geometric criteria, convergence cues, and feedback-driven control rules. In this vision, AI becomes not only a passive executor but an active collaborator, capable of refining algorithms by translating mathematical insights into precise, executable code adjustments.

Throughout the book, we will explore deterministic, geometry-based models that embody self-correcting behavior, inspired by reinforcement learning frameworks but grounded firmly in classical analysis and differential geometry. We delve into iterative schemes that steer points or bodies toward equilibrium states, dissect convergence properties, and examine the role of adaptive learning rates and control policies in shaping algorithmic trajectories.

While this work is intentionally abstract and does not presume immediate application to physical systems, it lays the groundwork for understanding how AI-driven direct programming can leverage rigorous mathematics to *intelligently guide* algorithmic development. The insights gleaned here, promise to inform future research bridging symbolic mathematical reasoning, AI interpretability, and computational geometry, opening new pathways for autonomous algorithm design.

It is important to emphasize that this book does **not** claim to be exhaustive. Instead, it embraces mathematical speculation and conceptual exploration, presenting models and ideas that are often abstract and intended to provoke thought rather than serve as definitive solutions. This approach invites readers to engage creatively with the material and to consider new possibilities in the theory of straightening and AI-driven program adaptation.

I. Geometry - considerations

I.1 Geometric Modeling — Manifold Representations

A body submitted to straightening can be modeled as a 3D embedded Riemannian manifold $M \subset \mathbb{R}^3$, with a parametric surface $X(u,v) \in \mathbb{R}^3$:

$$X(u,v) = \begin{bmatrix} x(u,v) \\ y(u,v) \\ z(u,v) \end{bmatrix}, (u,v) \in D \subset \mathbb{R}^2$$

Definition of $D \subset \mathbb{R}^2$: In differential geometry a parametric surface X(u,v) maps a 2D parameter space D (a domain in $\mathbb{R}^2$) to 3D space $\mathbb{R}^3$:

$$X: D \subset \mathbb{R}^2 \to \mathbb{R}^3$$

So, D is the parameter domain — a region in the UV-plane — over which the surface is defined.

- Differential quantities (like tangent vectors Xu, Xv , metric tensor, curvature) are computed on D.
- The finite element mesh in simulation is built over a discretized version of D.
- The Jacobian determinant $\left|\frac{\partial(X)}{\partial(u,v)}\right|$ is essential for integrals over surfaces (e.g., for energy, strain, or heat).

The first fundamental form encodes lengths and angles:

$$I = \begin{bmatrix} E & F \\ F & G \end{bmatrix}, with\ E = X_u \cdot X_u, F = X_u \cdot X_v, G = X_v \cdot X_v$$

The second fundamental form describes curvature:

$$II = \begin{bmatrix} e & f \\ f & g \end{bmatrix}, with\ e = X_{uu} \cdot N, f = X_{uv} \cdot N, g = X_{vv} \cdot N$$

N — The Unit Normal Vector

For a smooth parametric surface $X(u,v) \in \mathbb{R}^3$, the unit normal vector N(u,v) at each point is defined as:

$$N = \frac{X_u \times X_v}{||X_u \times X_v||}$$

Where:

$$X_u = \frac{\partial X}{\partial u}; X_v = \frac{\partial X}{\partial v}; X_u \times X_v: normal(not\ yet\ normalized); ||\cdot||: euclidian\ norm$$

Geometric Meaning

The vector N is:

- Orthogonal to the tangent plane spanned by Xu and Xv
- Oriented consistently across the surface (if possible)
- Used to measure how the surface curves in 3D space, via the second derivatives Xuu , Xuv, Xvv

Compute II from X(u,v) - compute the second fundamental form of the surface defined by the parametric function $X(u, v) \in \mathbb{R}^3$

Compute partial derivatives: Xu, Xv, Xuu, Xuv, Xvv

Compute the unit normal:

$$N = \frac{X_u \times X_v}{||X_u \times X_v||}$$

Dot second derivatives with the normal: e=Xuu·N, f=Xuv·N, g=Xvv·N

Assemble: $II = \begin{bmatrix} e & f \\ f & g \end{bmatrix}$

Using the first fundamental form I and the second II, you can compute principal curvatures by solving:

det(II−κI)=0 (differential geometry –characteristic equation used to find the principal curvatures κ_1 and κ_2 of a surface)

The roots κ1,κ2 are the principal curvatures, and:

Mean curvature $H = \frac{1}{2}(k_1 + k_2)$

Gaussian curvature $K = k_1 k_2 = \frac{detII}{detI}$

Curvature Evolution

We can track the change in curvature using shape operators and Gaussian curvature:

- Shape operator: S=−dN
- Mean curvature H, Gaussian curvature K:

$$H = \frac{eg - 2f^2 + ge}{2(EG - F^2)}, K = \frac{eg - f^2}{EG - F^2}$$

This allows tracking of localized warping and residual curvature.

I.2 Deformation Mapping

Let the body be deformed from configuration $X_{(u,v)} to \tilde{X}_{(u,v)}$.

Where $\tilde{X}_{(u,v)}$ is the deformed surface. In mechanics and differential geometry, especially when modeling deformations of surfaces (like during straightening):

Reference Configuration: $X(u, v) \in \mathbb{R}^3$ represents the initial (undeformed) surface geometry parameterized by$(u, v) \in D \subset \mathbb{R}^2$

Deformed Configuration: $\tilde{X}_{(u,v)} = X_{(u,v)} + u_{(u,v)}$

Where:

- u(u,v) is the displacement vector at point (u,v)
- $X(u, v) \in \mathbb{R}^3$describes the new position of the point after deformation

Example: Straightening Deformation

Imagine a warped beam with a deflection w(u,v)in the z-direction:

$$\tilde{X}_{(u,v)} = \begin{bmatrix} u \\ v \\ w_{(u,v)} \end{bmatrix}$$

$$\tilde{X}_{(u)} = \begin{bmatrix} 1 \\ 0 \\ \frac{\partial w}{\partial u} \end{bmatrix} \text{ and } \tilde{X}_{(v)} = \begin{bmatrix} 0 \\ 1 \\ \frac{\partial w}{\partial v} \end{bmatrix} \text{ and } \tilde{N} = \frac{\begin{bmatrix} \left(-\frac{\partial w}{\partial u}\right) \\ \left(-\frac{\partial w}{\partial v}\right) \\ 1 \end{bmatrix}}{\sqrt{1 + \left(\frac{\partial w}{\partial u}\right)^2 + \left(\frac{\partial w}{\partial v}\right)^2}}$$

rotation + translation of a 3D surface (see fig.1 foe a point) fully consistent with modeling a deformed surface:

$$x = x' cos\alpha_1 + y' cos\alpha_2 + z' cos\alpha_3$$

$$y = x' cos\beta_1 + y' cos\beta_2 + z' cos\beta_3$$

$$z = x' cos\gamma_1 + y' cos\gamma_2 + z' cos\gamma_3$$

Orthogonality

$$cos\alpha_1^2 + cos\beta_1^2 + cos\gamma_1^2 = 1$$

$$cos\alpha_2^2 + cos\beta_2^2 + cos\gamma_2^2 = 1$$

$$cos\alpha_3^2 + cos\beta_3^2 + cos\gamma_3^2 = 1$$

Normalizing

$$cos\alpha_1 cos\alpha_2 + cos\beta_1 cos\beta_2 + cos\gamma_1 cos\gamma_2 = 0$$

$$cos\alpha_2 cos\alpha_3 + cos\beta_2 cos\beta_3 + cos\gamma_2 cos\gamma_3 = 0$$

$$cos\alpha_3 cos\alpha_1 + cos\beta_3 cos\beta_1 + cos\gamma_3 cos\gamma_1 = 0$$

$$x = x' + a$$

$$y = y' + b$$

$$z = z' + c$$

Fig. 1: visualization of

$$\begin{bmatrix} x \\ y \\ z \end{bmatrix} = \begin{bmatrix} cos\alpha_1 & cos\alpha_2 & cos\alpha_3 \\ cos\beta_1 & cos\beta_2 & cos\beta_3 \\ cos\gamma_1 & cos\gamma_2 & cos\gamma_3 \end{bmatrix} \cdot \begin{bmatrix} x` \\ y` \\ z` \end{bmatrix} + \begin{bmatrix} a \\ b \\ c \end{bmatrix}$$

original coordinate system: basis vectors e1 = {1, 0, 0}   e2 = {0, 1, 0}   e3 = {0, 0, 1}

Translation vector: T = {2, 1, 1.5} and 45° rotation in xy-plane

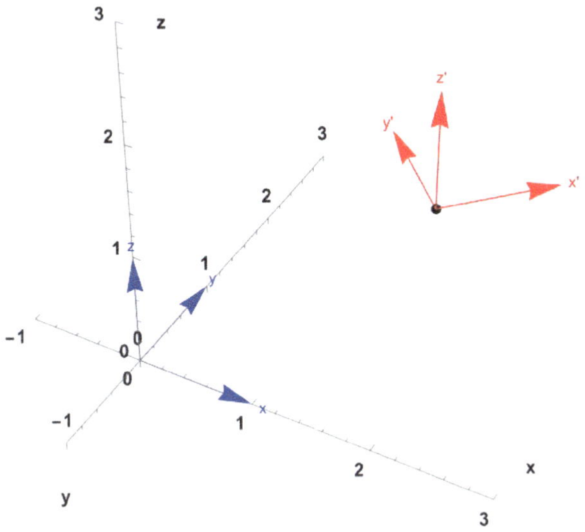

The deformation gradient F is:

$F_{(u,v)} = \nabla\tilde{X} = \frac{\partial\tilde{X}}{\partial u} \otimes e_u + \frac{\partial\tilde{X}}{\partial v} \otimes e_v$ this defines the deformation gradient tensor for a surface, expressed in the parametric coordinates where:

- $X(u,v) \in \mathbb{R}^3$ is the deformed position of each point originally parameterized by (u,v)
- $F(u,v) \in \mathbb{R}^{3x2}$ maps infinitesimal vectors in the (u,v) plane to vectors in 3D space

So, F gives the local linear transformation between the undeformed (reference) tangent plane and the deformed surface geometry.

- $\frac{\partial\tilde{X}}{\partial u}$ is a tangent vector in 3D space showing how the surface moves with u
- $\otimes$ denotes the tensor product
- e_u, e_v are unit vectors in the u and v directions in the 2D parameter domain

The deformation gradient F is central for:

Use Case	Quantity Derived From F
Strain analysis (linear or nonlinear)	Green–Lagrange strain: $E = \frac{1}{2}(F^T F - I)$
Curvature analysis	Used to compute how surface normals rotate under deformation
Finite Element Method (FEM)	Provides shape function gradients in physical space
Machine Learning control input	Encodes geometric distortion for AI-based tooling or correction systems

It's a quadratic measure of strain that:

- Works for large deformations
- Measures how much a material deforms relative to its original configuration
- Is zero if the deformation is a pure rigid motion (rotation + translation)

It captures both stretching and shearing.

- $F(u,v) \in \mathbb{R}^{3x2}$: deformation gradientfor a 2D surface embedded in 3D
- $I \in \mathbb{R}^{2x2}$: identity matrix in the parameter space
- $F^T F \in \mathbb{R}^{2x2}$: metric tensor of the deformed configuration

Tensor	Size	Meaning
F	3×2	Maps 2D parametric displacements to 3D space
F^T	2×3	Maps 3D vectors to parameter-space
$F^T F$	2×2	Pullback of metric (used in E)
$E = \frac{1}{2}(F^T F - I)$	2×2	Green–Lagrange strain on the surface

Physical Meaning in Surface Straightening

Body correction:

- E(u,v) tells where and how much the body has stretched, shrunk, or sheared
- The goal in straightening is often to minimize ||E(u,v)|| over the domain
- You can input E to a self-learning control system to identify patterns of warpage or correct tooling paths

T is the tangent space to the surface at a point — and it is: TpS≅ $\mathbb{R}^2$

That is, T is the 2D tangent plane (the parameter space), and all vectors you apply F to live in this 2D space.

Let:

- S be the smooth surface embedded in $\mathbb{R}^3$
- p=X(u,v)∈S be a point on that surface
- TpS: the tangent space to the surface at p

Then:

- T=TpS≅span{Xu, Xv}⊂ $\mathbb{R}^3$ - In linear algebra, the span of a set of vectors is the set of all linear combinations of those vectors; often compare tangent vectors before and after deformation: $span\{X_u, X_v\}\ vs.\ span\{\tilde{X}_{(u)}, \tilde{X}_{(v)}\}$

In parameter space (undeformed), the tangent vectors are: $\frac{\partial}{\partial u}, \frac{\partial}{\partial v}$. So, the tangent space is naturally isomorphic to $\mathbb{R}^2$, which is why:

- The deformation gradient F: $\mathbb{R}^2 \rightarrow \mathbb{R}^3$ **maps** parametric differentials into real-space vectors
- $F^T F$ operates within $\mathbb{R}^2$— measuring distortions within the tangent plane

II. Geometrical straightening effort - a speculative approach

To calculate the straightening effort required to transform a point $A_2(x_2, y_2, z_2)$ to $A_1(x_1, y_1, z_1)$ we first need to clarify what is meant by "straightening effort." Based on the earlier context (rotation + translation using direction cosines), we interpret this as:

This involves:

1. Calculating the translation vector.
2. Estimating the rotation matrix (based on direction cosines).
3. Finding the transformation parameters.
4. Computing the "effort," which can be quantified using:
   - Euclidean distance (for translation),
   - Rotation angle/matrix norm (for rotation),
   - Or a combined cost function (e.g., Frobenius norm).

Step 1: Translation Vector
Let's define:

$$\vec{t} = \begin{bmatrix} t_x \\ t_y \\ t_z \end{bmatrix} = \begin{bmatrix} x_1 - x_2 \\ y_1 - y_2 \\ z_1 - z_2 \end{bmatrix}$$

Step 2: Translation "Effort"
The translation effort can be quantified using Euclidean norm:

$$\|\vec{t}\| = \sqrt{(x_1 - x_2)^2 + (y_1 - y_2)^2 + (z_1 - z_2)^2}$$

Step 3: Rotation Matrix R
If the local coordinate frames at A_1 and A_2 are known (i.e., local axes before and after straightening), the rotation matrix R that aligns them can be calculated using:

- Procrustes analysis
- Kabsch algorithm
- Or simply based on direction cosines if already known.

Let: R=3×3 orthonormal rotation

Then the rotation effort can be estimated via rotation angle θ derived from:

$$\theta = cos^{-1}\left(\frac{T_r(R) - 1}{2}\right)$$

This gives a scalar measure of angular effort (in radians or degrees).

Step 4: Total Effort

One common way to combine translation and rotation effort is:

$$Total\ Effort = w_t \cdot \|\vec{t}\| + w_r \cdot \theta$$

Where:

- w_t , w_r are weights (tunable depending on application).
- E.g., $w_t = 1$, $w_r = 1$ if treated equally.

Example: straightening from A_2 to A_1: $A_1 = (1, 2, 3)$ $A_2 = (1.5, 2.5, 3.5)$

Translation vector: $\vec{t} = \begin{bmatrix} -0.5 \\ -0.5 \\ -0.5 \end{bmatrix}$ $and\ \|\vec{t}\| \approx 0.867$

Assumed Orientation (Rotation Matrices)

We define the local coordinate systems as rotation matrices, where each column is a local axis vector.

A1 orientation: aligned with global axes

$$R_1 = \begin{bmatrix} 1 & 0 & 0 \\ 0 & 1 & 0 \\ 0 & 0 & 1 \end{bmatrix}$$

A2 orientation: rotated 60° around the Y-axis

$$R_2 = \begin{bmatrix} \cos(60°) & 0 & \sin(60°) \\ 0 & 1 & 0 \\ -\sin(60°) & 0 & \cos(60°) \end{bmatrix} = \begin{bmatrix} 0.5 & 0 & -0.866 \\ 0 & 1 & 0 \\ 0.866 & 0 & 0.5 \end{bmatrix}$$

Compute Rotation Angle θ

$$\theta = cos^{-1}\left(\frac{T_r(R) - 1}{2}\right)$$

Trace of R: Tr(R)=0.5+1+0.5=2.0

Then:

$$\theta = cos^{-1}\left(\frac{2-1}{2}\right) = 60° = 1.047\ radians$$

Combine Efforts (based on Euclidean distance + 60° angular rotation)

Assuming equal weights for translation and rotation:

$$Total\ Straightening\ Effort = \|\vec{t}\| + \theta = 0.866 + 1.047 = 1.913\ conventional\ units\ of\ geometrical\ straightening\ effort$$

II.1 Geometrical effort based on the productivity definition of Productivity of a Domain (inspired by [1])

Let

$$\vec{F}: D \subset \mathbb{R}^3 \rightarrow \mathbb{R}^3$$

be a vector field of class C1 (i.e., continuously differentiable). We define the productivity of a domain $V \subset D$ as the quantity $\iiint_V div\vec{F}\ dV$ measured in conventional units. The author changed the productivity into effort considering this an appropriate mach. [1]

Compare how two points in 3D space influence a region around them, based on how far they are from every other point. This "influence" or effort is to be quantified as a scalar value, and the difference between the two efforts gives a sense of how "much more" one point affects the space than the other.

$$\vec{P}_1 = (x_1, y_1, z_1) \; and \; \vec{P}_2 = (x_2, y_2, z_2)$$

evaluate the influence throughout a region of space, using a variable point:

$$\vec{r} = (x, y, z)$$

This is the integration variable, sweeping through a 3D domain (a volume V).

Using the adapted definition of effort (see above productivity → effort we obtain:

$$\int_{xmin}^{xmax} \int_{ymin}^{ymax} \int_{zmin}^{zmax} div\left(\sqrt{(x_1 - x_0)^2 + (y_1 - y_0)^2 + (z_1 - z_0)^2}\right) dxdydz$$

$$\int_{xmin}^{xmax} \int_{ymin}^{ymax} \int_{zmin}^{zmax} div\left(\sqrt{(x_2 - x_0)^2 + (y_2 - y_0)^2 + (z_2 - z_0)^2}\right) dxdydz$$

Generic

$$\iiint_V \nabla \cdot f(\vec{r}) dV$$

where $f(\vec{r}) = \|\vec{P}_i - \vec{r}\|$, the distance between $\vec{P}_i$ and $\vec{r}$

But this is mathematically ill-defined because:

- the divergence of a scalar isn't meaningful.
- Divergence applies to vector fields, not scalar fields.

Correct Formulation Using Vector Fields

Instead of scalar distance functions, now consider:

$$\vec{F}_i(\vec{r}) = \frac{\vec{r} - \vec{P}_i}{\|\vec{r} - \vec{P}_i\|^3}$$

This is a classical inverse-square vector field, pointing away from $\vec{P}_i$ and decaying with distance.

This field is:

- Analogous to gravitational or electrostatic fields
- Has divergence equal to zero everywhere, except at the point $\vec{P}_i$.

At $\vec{P}_i$ the divergence is a Dirac delta function:

$$\nabla \cdot \vec{F}_i(\vec{r}) = 4\pi\delta(\vec{r} - \vec{P}_i)$$

This means:

- The field has no net divergence (no creation or loss of "flow") anywhere else
- All the divergence is concentrated at the point source

The symbol δ here represents the Dirac (see fig.1) delta function, and $\delta(\vec{r} - \vec{P}_i)$ is the 3D Dirac delta function centered at point $\vec{P}_i$.

Key properties:

- Not a function in the classical sense, but a generalized function or distribution.
- It is zero everywhere except at $\vec{r} = \vec{P}_i$
- Its integral over space equals 1:
- $\iiint_{\mathbb{R}^3} \delta(\vec{r} - \vec{P}_i) dV = 1$
- A "point mass" or infinitely sharp spike at $\vec{P}_i$

Fig.1

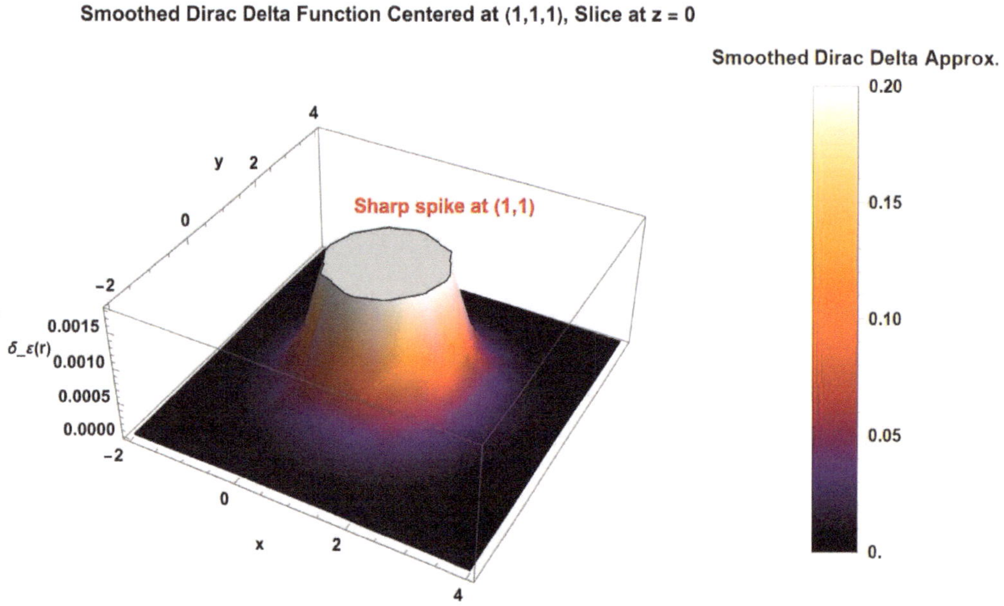

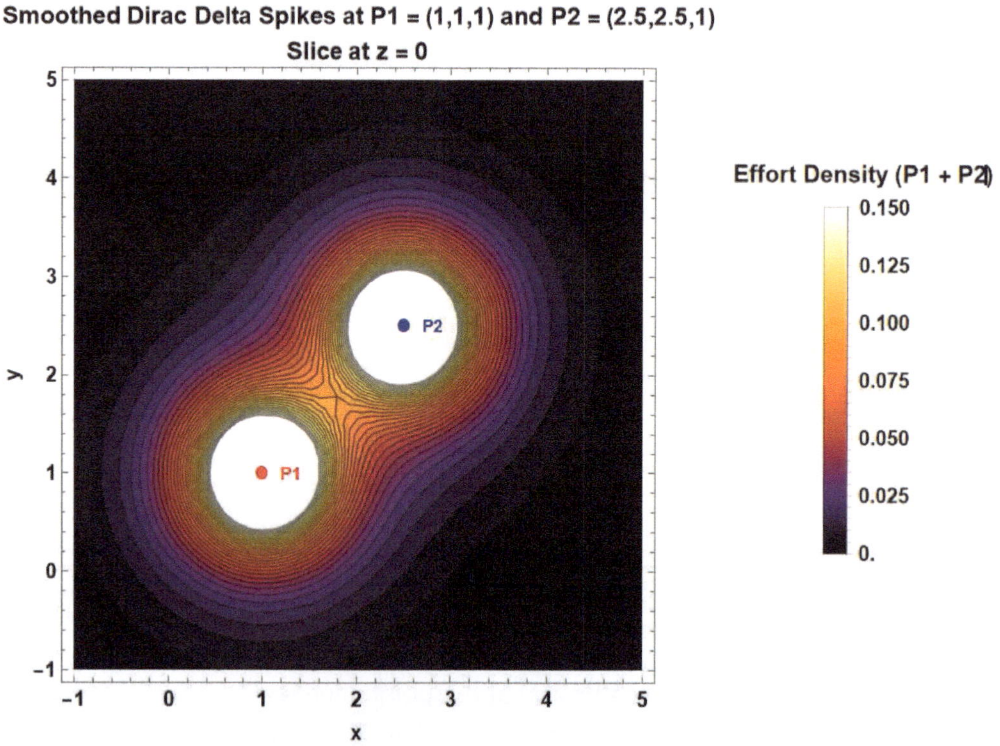

Interpretation

- Each spike is a localized area of influence
- Overlap of the contours indicates regions affected by both
- One can use this setup to visualize competition or combined influence between two sources

So the total influence/effort over a region V is:

$$E_i = \iiint_V \nabla \cdot \vec{F}_i \vec{r} dV = \begin{cases} 4\pi, if\ \vec{P}_i \in V \\ 0, otherwise \end{cases}$$

This gives a sharp "all or nothing" result — not helpful if $\vec{P}_i$ is near the edge or outside.

Smooth Approximation: Effort Density

To address this, you define a smooth scalar function that approximates how strong the influence of$\vec{P}_i$ is at any location $\vec{r}$:

$$\vec{D}_i(\vec{r}) = \frac{\varepsilon^2}{\left\|\vec{r} - \vec{P}_i\right\|^5 + \varepsilon^2}$$

Each term explained:

- $\left\|\vec{r} - \vec{P}_i\right\|$: Distance between the evaluation point and the source point
- Power 5: Chosen because it gives rapid decay and smooth integrability
- ε^2 : A small regularization constant to avoid division by zero when $\vec{r} = \vec{P}_i$

This function:

- Peaks sharply at $\vec{r} = \vec{P}_i$
- Decays quickly with distance
- Is smooth and numerically stable
- Mimics the effect of the delta function from the original divergence formulation

Total Effort of Each Point

To quantify how much each point affects a region V, integrate this scalar density over that region:

$$E_i = \iiint_V D_i(\vec{r}) dV$$

This scalar value represents the total "effort" from point $\vec{P}_i$ over the region V.

Effort Difference

To compare how much more one point influences the region than the other:

ΔE=E1−E2

Where:

- E_1 effort from $\vec{P}_1$
- E_2 effort from $\vec{P}_2$

So ΔE is the net difference in influence of the two points across the same space.

Summary Table

Term	Meaning
$\vec{P}_1, \vec{P}_2$	Source points of influence
$\vec{r} = (x, y, z)$	Variable point in space
$\|\vec{r} - \vec{P}_i\|$	Distance from evaluation point to source
$\vec{F}_i(\vec{r})$	Inverse-square vector field
$\nabla \cdot \vec{F}_i$	Divergence =Dirac delta
$D_i(\vec{r})$	Smooth scalar „effort density"
$E_i = \iiint_V D_i(\vec{r})dV$	Total effort from point $\vec{P}_i$
ΔE=E1−E2	Effort difference

To visualize (see fig. 2) the scalar field:

- Slice the 3D region at z=0
- Plot the scalar field D1(x,y,0), D2(x,y,0) and their difference
- Use color maps to show how the effort spreads across space
- Effort at any point (x,y,0) due to a source located at P1 (or P2) — i.e., effort as a function of the distance between P1 (or P2) and the point (x,y,0).

Fig.2 Visualization

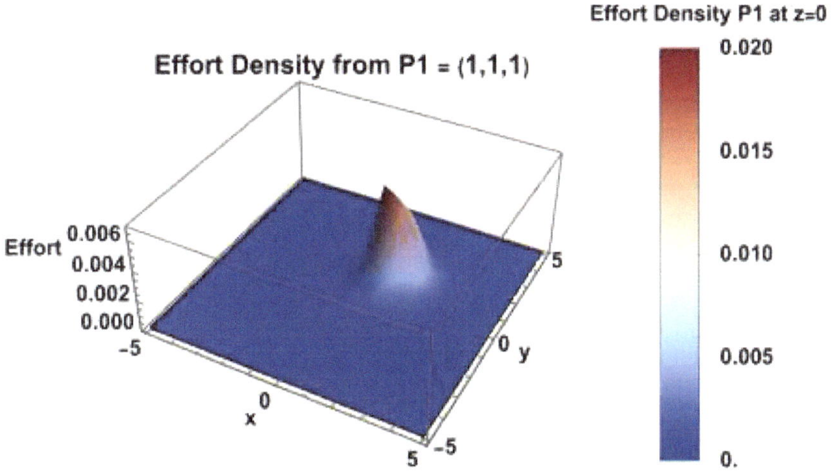

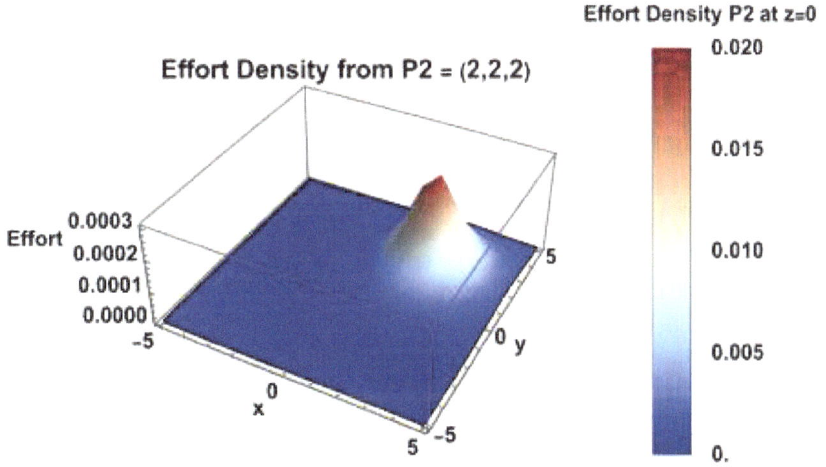

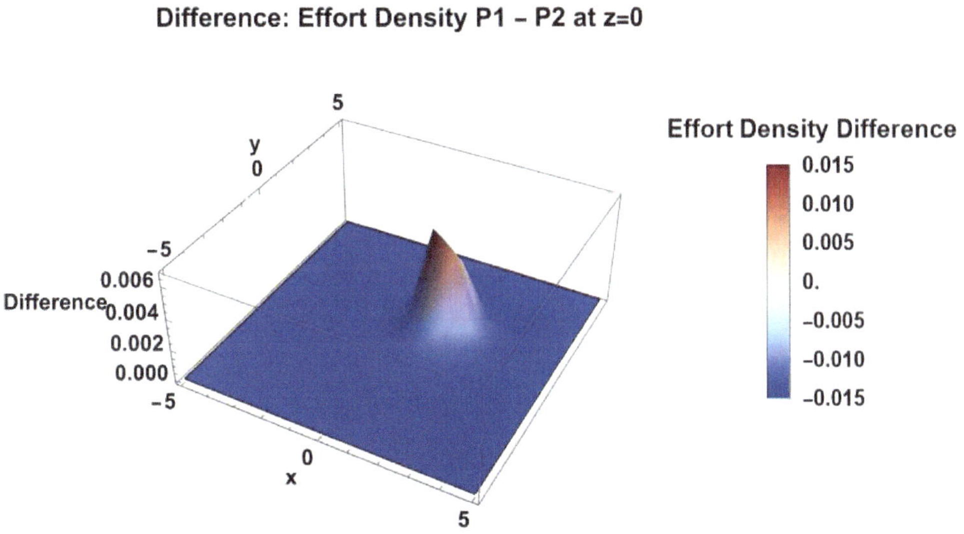

III. Mechanical straightening effort - a simplified speculative approach

Mechanical Effort

This refers to actual physical effort: forces, torques, energy, or work required to perform the straightening.

It depends on:

- Material properties (stiffness, elasticity, damping)
- Constraints and supports
- Force distribution
- Load path and deformation behavior

In other words:

Mechanical Effort~Forces and Moments needed to produce the movement

There is a relationship between geometrical and mechanical effort, especially in elastic or rigid bodies:

Simplified View:

Mechanical Effort= Stiffness × Geometrical Effort

Or more precisely:

- Force is proportional to displacement (Hooke's law):$\vec{F} = K \cdot \vec{u}$
- Moment (torque) is proportional to rotation: $\vec{M} = C \cdot \vec{\theta}$

  Where:
- K is the translational stiffness matrix
- C is the rotational stiffness matrix
- $\vec{u}$ is translation (displacement)
- $\vec{\theta}$ is rotation vector

Important Notes

- Linear relation only holds for small deformations or linear elastic materials.
- For nonlinear materials, friction, or plastic deformation, the relation becomes more complex.
- The distribution of deformation also matters — e.g., a long flexible beam may need less force for the same geometrical effort compared to a short, stiff beam.

Mechanical Effort

- F is the force required to achieve translation u, depending on stiffness matrix K
- M is the torque (moment) required to rotate by θ, depending on rotational stiffness matrix C
- Emech is the total mechanical effort (strain energy) needed to straighten the point

Let:

$$\vec{u} = \begin{bmatrix} u_x \\ u_y \\ u_z \end{bmatrix} = \begin{bmatrix} x_i - x_1 \\ y_i - y_1 \\ z_i - z_1 \end{bmatrix} \text{ and } \vec{\theta} = \begin{bmatrix} \theta_x \\ \theta_y \\ \theta_z \end{bmatrix} (rottion\ vector)$$

Stiffness matrices:

$$K = \begin{bmatrix} k_{11} & k_{12} & k_{13} \\ k_{12} & k_{22} & k_{23} \\ k_{13} & k_{23} & k_{33} \end{bmatrix} \ and\ C = \begin{bmatrix} c_{11} & c_{12} & c_{13} \\ c_{12} & c_{22} & c_{23} \\ c_{13} & c_{23} & c_{33} \end{bmatrix}$$

Each element k_{ij} means:

The force in direction i due to a unit displacement in direction j.

Examples:

- k_{11} : force in x caused by x-displacement → axial stiffness in x
- k_{23} : force in y caused by z-displacement → shear-like coupling
- k_{12}: force in x caused by y-displacement → cross-axis coupling

Each element c_{ij} means:

The moment (torque) in direction i due to a unit rotation in direction **j.**

Examples:

- c_{11} : moment in x due to rotation about x → torsional stiffness
- c_{12} : moment in x due to rotation about y → torsion-bending coupling
- c_{33} : stiffness resisting twisting about z

Term	Meaning
k_{ij}	Force in direction ii due to translation in direction j
c_{ij}	Moment (torque) in direction ii due to rotation in direction j
Diagonal	Direct stiffness (along/around same axis)
Off-diagonal	Coupling between axes (shear, torsion-bending, etc.)

Important Notes

- Both K and C must be symmetric and positive semi-definite for physical realism (no negative energy).
- If the system is orthotropic ("Same in all directions" — mechanical properties are identical in every direction), some off-diagonal terms may be non-zero.
- If the system is isotropic ("Different in 3 orthogonal directions" — properties vary along x, y, and z axes but are symmetric with respect to those axes), the matrices are diagonal.

Force Vector (from translation)

$$\vec{F} = K \cdot \vec{u} = \begin{bmatrix} k_{11}u_x + k_{12}u_y + k_{13}u_z \\ k_{12}u_x + k_{22}u_y + k_{23}u_z \\ k_{13}u_x + k_{12}u_y + k_{33}u_z \end{bmatrix}$$

Moment Vector (from rotation)

$$\vec{M} = C \cdot \overrightarrow{\theta\theta} = \begin{bmatrix} c_{11}\theta_x + c_{12}\theta_y + c_{13}\theta_z \\ c_{12}\theta_x + c_{22}\theta_y + c_{23}\theta_z \\ c_{13}\theta_x + c_{12}\theta_y + c_{33}\theta_z \end{bmatrix}$$

Total Mechanical Effort (Strain Energy)

This represents the mechanical work required to apply the translation and rotation (assuming linear elasticity):

$$E_{mech} = \frac{1}{2}\vec{u}^T K \vec{u} + \frac{1}{2}\vec{\theta}^T C \overrightarrow{\theta}$$

Where:

- $\vec{u}$ translation vector (units: m)
- $\overrightarrow{\theta}$ rotation vector (units: rad)
- K translational stiffness matrix (units: N/m)
- C rotational stiffness matrix (units: N·m/rad or N·m)
- Emech mechanical effort / strain energy (units: Joules = N·m)

Term	Expression	Units
Translational effort	$\vec{u}^T K \vec{u}$	N·m = J
Rotational effort	$\vec{\theta}^T C \overrightarrow{\theta}$	N·m = J
Total mechanical energy	E_{mech}	Joules

Expanded form:

$$E_{mech} = \frac{1}{2}\left[k_{11}u_x^2 + 2k_{12}u_x u_y + 2k_{13}u_x u_z + k_{22}u_y^2 + 2k_{23}u_y u_z + k_{33}u_z^2\right] + \frac{1}{2}\left[c_{11}\theta_x^2 + 2c_{12}\theta_x\theta_y + 2c_{13}\theta_x\theta_z + c_{22}\theta_y^2 + 2c_{23}\theta_y\theta_z + c_{33}\theta_z^2\right]$$

Symbol	Description	Example Value(see fig. 1)	Unit
k	Translational stiffness	1000	N/m
c	Rotational stiffness	200	N·m/rad
$\vec{u}$	Translation vector (e.g. [x,0,0])	0 to 0.03	meters
$\vec{\theta}$	Rotation vector (e.g. [θ,0,0])	0 to π/6	radians (° for UI)
E_{mech}	Mechanical effort (strain energy)	computed	Joules (N·m)

Fig.1 Mechanical Effort example

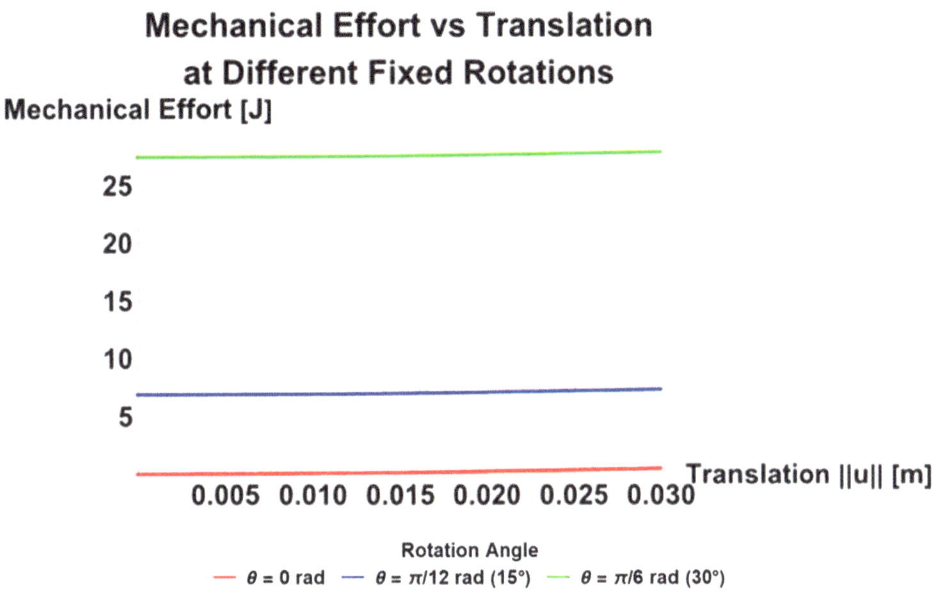

IV. Differential equations of motion - a simplified approach

Elasticity Model: Nonlinear Elasto-Plasticity

Let u(x,t) be the displacement field. The balance of linear momentum in a deformable body is:

$$\rho \frac{\partial^2 u}{\partial t^2} = \nabla \cdot \partial + f$$

with:

- σ: Cauchy stress tensor
- f: body forces
- ρ: density

The constitutive law (body = elastoplastic) could be: $\sigma = \mathbb{C}: (\varepsilon - \varepsilon^p)$ central for modeling deformation such as body straightening.

This is the constitutive law (material law) for elastoplasticity in small strains, sometimes called the additive decomposition model.

- σ: Cauchy stress tensor (symmetric 2nd-order tensor)
- ε: total strain tensor (usually Green–Lagrange for large strains, but infinitesimal here)
- ε^p: plastic strain tensor (irreversible)
- ℂ: 4th-order elasticity tensor — it defines the linear elastic material properties (Hooke's law)
- : Double contraction, meaning a tensor product with two summed indices:

$$(\mathbb{C}: A)_{ij=\sum_{k,l} C_{ijkl} A_{kl}}$$

Index	Meaning
i,j	Output indices → components of the resulting 2nd-order tensor (the stress σ_{ij})
k,l	Summation indices → components of the input strain tensor ε_{kl}

ℂ is the 4th-order elasticity tensor:$C_{ijkl} \in \mathbb{R}^{3x3x3x3}$; in 3D it defines material stiffness: how stress σ_{ij} responds to strain ε_{kl}

The double contraction ℂ:ε computes:

$$\sigma_{ij} = \sum_{k=1}^{3} \sum_{l=1}^{3} C_{ijkl} \varepsilon_{kl}$$

For every i,j sum over all k,l → this gives one stress component σ_{ij}

Symbol	Meaning
C_{ijkl}	4th-order elasticity tensor (stiffness)
A_{kl}	Input 2nd-order tensor (strain)
$(C:A)_{ij}$	Output 2nd-order tensor (stress)
i,j	Output indices
k,l	Summation (dummy) indices

$$\sigma = \mathbb{C}:(\varepsilon - \varepsilon^p)$$

$$\varepsilon = \frac{1}{2}(\nabla u + \nabla u^T), \dot{\varepsilon}^p = \lambda \frac{\partial f}{\partial \sigma}$$

where:

ε^p: plastic strain

$f(\sigma)$: yield function (e. g. Von Mieses)

Λ: plastic multiplier (Karush-Kuhn-Tucker conditions apply)

Optional Additions

- Kuhn–Tucker conditions:

  $f \leq 0, \Lambda \geq 0, \Lambda f = 0$

- These ensure that plastic flow only occurs when the yield criterion is met.
- Hardening laws or rate-dependence can be added for more realism.

Hardening Laws

Hardening laws describe how the material's yield surface evolves with continued plastic deformation.

Why it's needed:

In reality, once a material yields, it doesn't just flow indefinitely with no added resistance. It often becomes stronger (harder) as it plastically deforms.

Types of hardening

Type	Concept	
Isotropic Hardening	Yield surface expands uniformly in stress space	$f(\sigma,\alpha) = \|dev(\sigma)\| - \sigma_y(\alpha)$
Kinematic Hardening	Yield surface translates in stress space	$f(\sigma - X) = \|dev(\sigma - X)\| - \sigma_y$e
Mixed Hardening	Combines both isotropic and kinematic effects	Depends on both α and X evolution

State Variables Introduced:

- α: Scalar internal variable (accumulated plastic strain)
- X: Backstress tensor (for kinematic hardening)
- Evolution laws like:

$$\dot{\alpha} = \Lambda, X = C\dot{\epsilon}^{p} - DX$$

Where

$\dot{\alpha}$	-	Rate of accumulated plastic strain (scalar internal variable)
Λ	-	Plastic multiplier (from flow rule, always ≥ 0)
$\dot{X}$	-	Rate of evolution of backstress (for kinematic hardening)
C	-	Kinematic hardening modulus (material parameter)
D	-	Recovery coefficient (defines how fast backstress decays)
$\dot{\epsilon}$	-	Plastic strain rate tensor

Rate-Dependence (Viscoplasticity)

Viscoplastic models account for time-dependent plastic flow — important for polymers, metals at high temperature, or very fast deformation. Example: Perzyna-Type Viscoplasticity

Why it's needed: Some materials deform plastically only when loaded fast enough, or show strain-rate sensitivity.

Dynamics of Deformation

- When deformation evolves over time (e.g. during straightening, vibration, impact), we must account for how motion and forces interact.
- Even if the final state is static, the path to that state (transient response) involves motion

V. Strategy: mechanically and computationally complex problem of *automatic straightening of a body* (such as a deformed beam, rod, or structure), where a self-learning algorithm (e.g., reinforcement learning or neural network) applies variable forces or impulses at specific points to return the structure to a desired geometry (e.g., straight or minimal strain configuration).

Let's consider:

- A body (e.g., a 1D rod, 2D plate, or 3D structure), discretized into n nodes or points: {p1,p2,…,pn}
- determine the optimal sequence, magnitude, and location of applied forces or impulses to drive the body to a straight configuration
- The physical response of the body is governed by continuum mechanics (e.g., elasticity, dynamics), and thus PDEs
- You aim to automate the process using a self-learning algorithm (e.g., reinforcement learning, deep neural control, etc.)

V.1 Mathematical Modeling of the Body

We'll start from physics — then link that to control and learning.

V.1.1 Governing Physical Equations

- Assume a deformable elastic body in 3D, with displacement field:

$$u(x,t) = x(t) - x_0$$

Where $x(t) \in \mathbb{R}^3$ is the current position of material point and $x_0 \in \mathbb{R}^3$ is the reference position

Dynamic Elasticity (Linear or Nonlinear):

$$\rho \frac{\partial^2 u}{\partial t^2} = \nabla \cdot \sigma + f_{ext}$$

- ρ = mass density
- σ = Cauchy stress tensor (function of strain and material model)
- f_{ext} = external force field (your applied impulses/forces)

V.1.2 Constitutive Law

For linear elasticity:

$$\sigma = C : \epsilon, \epsilon = \frac{1}{2}(\nabla \mathrm{u} + \nabla u^T)$$

Where: T denotes the transpose operator

where C is the 4th-order elasticity tensor (depends on Young's modulus, Poisson ratio, etc.)

For nonlinear behavior, use hyperelastic models (e.g., Neo-Hookean, Ogden, Mooney-Rivlin).

Impulses or Forces:

$$f_{ext}(x,t) = \sum_{i=1}^{k} F_i\,(t)\delta(x - x_i)$$

$F_i(t)$: time-varying force applied at location x_i

δ(·): Dirac delta (impulse at point)

discretize the domain into finite elements or nodes, and write this as a second-order system:

V.2 Discretized Equation (Finite Element / FEM Form)

The continuous PDE leads to the discretized matrix form:

$$(M)(\ddot{u}) + (C)(\dot{u}) + (K)(u) = F_{ext}(t)$$

Where:

- M: Mass matrix
- C: Damping matrix
- K: Stiffness matrix
- $u \in \mathbb{R}^{3n}$: Displacement vector (each node has 3 DOF = Degrees of Freedom)
- $F_{ext} \in \mathbb{R}^{3n}$: External force vector (nonzero only at k nodes)

V.3 Self-Learning Control Strategy

- Decide which points to apply force to
- Determine magnitude and direction of each force
- Adapt this over time based on the response of the system

This is a control & optimization problem, with possible machine learning integration.

V.3.1 Objective Function

Define a loss or energy to minimize:

$$L = \int_0^T \left(\left\|u(t) - u_{target}\right\|^2 + \lambda\|F_{ext}(t)\|^2\right) dt$$

Or for a static case:

$$L = \left\|u_{final} - u_{target}\right\|^2 + \lambda \sum_{i=1}^{k} \|F_i\|^2$$

B) Learning Framework:

Use Reinforcement Learning (RL) or Neural ODE (Ordinary Differential Equations) control:

- State: current deformation u(t)
- Action: vector of forces $F_{1:k}$ to apply

- Reward: negative of L
- Environment: governed by the above differential equations

Tools:

- Use Physics-Informed Neural Networks (PINNs) to learn solutions to the PDE while enforcing physical laws
- Or use Model Predictive Control (MPC) with learned system dynamics
- Use Graph Neural Networks to encode the structure and dynamics of the mesh

V.4 Summary of Key Concepts

PDE Governing Equation	$\rho \ddot{u} = \nabla \cdot \sigma + f\_ext$
Constitutive Law	$\sigma = C : \varepsilon$ (linear or nonlinear)
Discrete FEM System	$M\ddot{u} + C\dot{u} + Ku = F$
Control Problem	Find sequence of Fi(t) to minimize distance to target
Learning Component	RL agent or neural controller selects and adapts forces
Objective Function	Combine deformation error + force penalty

V.5 Possible Mathematical Model of Mechanical Straightening

V.5.1 Geometric and Physical Setting

Let the body occupy a domain:

$$\Omega \in \mathbb{R}^d, d = 1, 2 \text{ } or \text{ } 3$$

with boundary $\partial\Omega$, and time domain $t\in[0,T]$.

Each material point in the body is indexed by its reference (initial) position $X\in \Omega$. The body's deformation is described by: $u(x,t): \Omega \times [0,T] \rightarrow \mathbb{R}^d$ which gives the **displacement field**: current position = $x=X+u(X,t)$.

V.5.2 Governing Equation (Elastic Dynamics with External Forces)

Newton's second law in continuum form:

$\rho \frac{\partial^2 u}{\partial t^2} = \nabla \cdot \sigma + f_{ext}(X,t)$

Where:

- ρ: mass density (scalar)
- σ: Cauchy stress tensor (depends on strain, material properties)
- fext: externally applied body force (control)

V.5.3 Material Law (Linear Elasticity)

- For isotropic, homogeneous, linear elastic material:

$$\sigma = \lambda tr(\epsilon)I + 2\mu\epsilon$$

tr(·) means trace of a matrix.

I is the identity tensor (or identity matrix).

$$\epsilon = \frac{1}{2}(\nabla u + (\nabla u^T))$$

Where:

- ε: linearized strain tensor
- λ,μ: Lamé parameters (related to Young's modulus E and Poisson ratio v)

V.5.4 Boundary and Initial Conditions

- Initial Conditions: $u(X,0) = u_0(X), \frac{\frac{\partial u}{\partial t}(X,0)u}{\partial t}(X,0) = v_o(X)$
- Meaning of Each Symbol

Symbol	Meaning
u(X,t)	Displacement vector field — how each material point X moves over time
$\frac{\partial u}{\partial t}(\boldsymbol{X}, \boldsymbol{t})$	Velocity — time derivative of displacement
X∈Ω	Initial (reference) position in the material body
t=0	Initial time
$v_0(X)$	Initial velocity field — a vector function giving the starting velocity at each point in the body

Boundary Conditions:

$u(X,t) = 0, for\ X \in \Gamma_D \subseteq \partial\Omega$ (D/Dirichlet, fixed)

$\sigma \cdot n = t_N(X,t) for\ X \in \Gamma_N = \frac{\partial\Omega}{\Gamma_D}$ (N/Neumann, surface traction)

V.6 Control: Applied Forces or Impulses

apply external forces at discrete or distributed points:

$f_{ext}(X,t) = \sum_{i=1}^{k(t)} F_i\,(t)\delta(X - X_i)$

Where:

- $F_i(t) \in \mathbb{R}^d$: force or impulse vector applied at point X_i
- k(t): number of points receiving force at time t
- δ: Dirac delta function

If using smoother approximations (like distributed forces), replace δ with a smooth mollifier.

V.7 Objective Functional (Straightening Goal)

Define an objective functional to minimize deformation from a desired "straight" configuration:

$$J[u,\{F_i\}] = \int_0^T \left(\int_\Omega \left\| u(X,t) - u_{target}(X) \right\|^2 dx + \lambda \sum_{i=1}^{k(t)} \|F_i(t)\|^2 \right) dt$$

Or in static final-time-only form:

$$J[u(T)] = \int_\Omega \left\| u(X,t) - u_{target}(X) \right\|^2 dx$$

- u_{target}: desired (e.g., straightened) configuration
- λ>0: regularization weight to penalize force magnitude

V.8 Optimization Problem (Control Formulation)

Minimize the objective J
Subject to the PDE constraints:

$\rho \frac{\partial^2 u}{\partial t^2}=\nabla\cdot\sigma+f_{ext}(X,t)$

And initial/boundary conditions.

V.9 Abstract Variational Formulation (Weak Form)

Let V be the appropriate Hilbert space $e.g. H^1(\Omega)$, then the weak form is:

$$Find\ u(t) \in V such\ that\ \forall v \in V:$$

$$\int_\Omega \rho \frac{\partial^2 u}{\partial t^2} \cdot v dX + \int_\Omega \sigma(u) : \nabla v dX = \int_\Omega f_{ext} \cdot v dX + \int_{\Gamma_N} t_N \cdot v dS$$

ρ: material density (mass per unit volume)

$\frac{\partial^2 u}{\partial t^2}$: acceleration of displacement field (i.e., physical acceleration)

σ(u): stress tensor as a function of displacement u

v: a test function (virtual displacement)

∇v: gradient of the test function

: double contraction (tensor inner product):

Γ_N: Neumann boundary (part of ∂Ω)

t_N : prescribed surface traction (force per unit area)

dS: surface element on the boundary Γ_N

In Lagrangian (material) coordinates:

- The body occupies a domain Ω in its initial state.
- Each point in Ω is labeled by a vector X∈Ω.
- Over time, the point originally at X moves to a new position
- x: current (deformed) position
- u(X,t) displacement vector from initial to current configuration

Summary Table of Terms

Term	Meaning	Physical Role
$\int_\Omega \rho\ddot{u} \cdot v\, dX$	Inertial forces	Mass × acceleration
$\int_\Omega \sigma(u) : \nabla v\, dX$	Internal (elastic) forces	From stress–strain
$\int_\Omega f_{ext} \cdot v\, dX$	Body forces	Like gravity
$\int_{\Gamma N} t_N \cdot v\, dS$	Surface forces	External loads

V.10 Optional: Discrete Form (Finite-Dimensional)

After FEM discretization:

$$(M)(\ddot{u}(t)) + (K)(u(t)) = F(t)$$

Where:

- M: mass matrix
- K: stiffness matrix
- F(t) force vector (sparse, only k entries non-zero at control points)

Then apply optimal control or reinforcement learning to determine time-varying $F(t) \in \mathbb{R}^{dn}$

Summary

Pure Mathematical Model

Domain: $\Omega \subset \mathbb{R}^d$

State variable: u(X,t)

Governing PDE: $\rho\frac{\partial^2 u}{\partial t^2} = \nabla \cdot \sigma + \sum_{i=1}^{k(t)} F_i\,(t)\delta(X - X_i)$

Stress-strain: $\sigma = \lambda tr(\epsilon)I + 2\mu\epsilon \quad \epsilon = \frac{1}{2}\left(\nabla u + (\nabla u^T)\right)$

Objective:

$$J[u, \{F_i\}] = \int_0^T \left(\int_\Omega \|u(X,t) - u_{target}(X)\|^2\, dx + \lambda \sum_{i=1}^{k(t)} ||F_i(t)||^2 \right) dt$$

Optimization: Minimize J subject to PDE, initial/boundary conditions

VI. Mathematical Model: Learning Algorithm for Mechanical Straightening

VI.1 State Equation (Physics)

Let:

$u(t) \subset \mathbb{R}^{dn}$: displacement vector of all points

$v(t) = \dot{u}(t) \in \mathbb{R}^{dn}$: velocity

$F(t, \theta) \in \mathbb{R}^{dn}$: control force, parameterized by learnable parameters $\theta \in \mathbb{R}^{p}$

θ: a vector of parameters

$\mathbb{R}^{p}$: the p-dimensional real vector space

- That means θ is a vector with p real-valued components.

The governing dynamics are:

$$M\ddot{u}(t) + Ku(t) = F(t;\theta)$$

Equivalently, define the state: $x(t) := \begin{bmatrix} u(t) \\ \dot{u}(t) \end{bmatrix} \in \mathbb{R}^{2dn}$

Then the state dynamics become:

$$\dot{x}(t) = \underbrace{\begin{bmatrix} \dot{u}(t) \\ M^{-1}(-Ku(t) + F(t;\theta)) \end{bmatrix}}_{:=f(x(t),\theta)}$$

VI.2 Objective Function (Cost / Loss): aim to reach a desired final state u_{target} , while minimizing effort. The cost is:

$$J(\theta) = \int_0^T \underbrace{\left\|u(t) - u_{target}\right\|^2 + \lambda\|F(t;\theta)\|^2)dt}_{:=L(x(t),\theta)}$$

Goal:

$$\underbrace{min}_{\theta} = J(\theta) = \int_0^T L(x(t),\theta)dt$$

Subject to:

$$\dot{x}\ (t) = f(x(t),\theta), x(0) = x_0$$

VI.3 Learning Dynamics (Gradient Descent)

We optimize θ via gradient-based learning, e.g., using the adjoint method or backpropagation through time (BPTT).

Let θ(s) evolve in fictitious "learning time" s, then:

$$\frac{d\theta}{ds} = -\nabla_\theta J(\theta)$$

This defines a gradient flow in parameter space:$\dot{\theta} = -\nabla_\theta \int_0^T L(x(t),\theta)dt$

This can be solved using variational calculus or automatic differentiation.

VI.4 Overall Learning-Controlled System

Let's combine the full system of equations:

Forward Dynamics: $\dot{x} = f(x(t), \theta), x(0) = x_0$

Objective (Functional to Minimize):$J(\theta) = \int_0^T \underbrace{\|u(t) - u_{target}\|^2 + \lambda\|F(t;\theta)\|^2)dt}_{:=L(x(t),\theta)}$

Parameter Learning:$\dot{\theta} = -\nabla_\theta J(\theta)$

VI.5. Extensions (Reinforcement Learning or Optimal Control)

If one prefers reinforcement learning (RL) instead of gradient control:

State: x(t)

Action: a(t)=F(t)

Policy: $\pi_0: x(t) \mapsto a(t)$

Reward: $r(t) = -L(x(t), \pi_0(x))$

Then optimize expected cumulative reward: $\underbrace{max}_{\theta}\ \mathbb{E}\left[\int_0^T r(t)dt\right]$

Symbol	Meaning
$\theta \in \mathbb{R}^p$	Parameter vector to be optimized (e.g. control inputs, policy parameters, force application strategy)
r(t)	Instantaneous reward (or negative cost) at time t — depends on system state, control, and possibly θ
$\int_0^T r(t)\,dt$	Cumulative reward over a time interval [0, T]
$\mathcal{E}[\cdot]$	Expectation over uncertainty (e.g., noise, initial conditions, randomness in dynamics)
$\max_\theta$	We are choosing parameters θ that maximize expected reward

This structure supports:

- Physics-based control
- Learning-adapted force application
- Real-time feedback-based training

VI.6 Why Gradient Is Noisy in RL

In RL, the objective is often:$J(\theta) = \mathbb{E}_{\tau\sim\pi_\theta} \sum_{t=0}^{T} r(t)$

π_θ: a policy (decision rule) parameterized by θ

τ: a trajectory (sequence of states and actions)

The expectation $\mathbb{E}$ is taken over random trajectories due to:

- Stochastic environment transitions
- Stochastic policies (e.g., sampling actions)

So, computing $\nabla_\theta J$ relies on sampling many trajectories, and therefore the gradient estimate is noisy.

Example: Policy Gradient

In policy gradient RL, the update rule is: $\theta^{k+1} = \theta^k + \alpha \nabla_\theta J(\theta)$

But:$\nabla_\theta J(\theta) \approx \frac{1}{N}\sum_{i=1}^{N}\sum_{t=0}^{T} \nabla_\theta log\pi_\theta(a_t^i \mid s_t^i) \cdot R_t^i$

Symbol	Description
N	Number of sampled episodes or trajectories
T	Time horizon (number of steps per trajectory)
a_t^i	Action taken at time t in the i-th trajectory
s_t^i	State at time t in the i-th trajectory
$\pi_\theta(a_t^i \mid s_t^i)$	Policy: probability of taking action a_t given state s_t, parameterized by θ
$log\pi_\theta(a_t^i \mid s_t^i)$	Log-likelihood of the action under the current policy
$\nabla_\theta log\pi_\theta(a_t^i \mid s_t^i)$	Gradient of the log-likelihood w.r.t. parameters — tells how the action probability changes as θ changes
R_t^i	Cumulative reward from time t onward in trajectory i (i.e., return), often:

$R_t^i = \sum_{k=t}^{T} \gamma^{k-t} r_k^i$ where $\gamma \in [0,1]$ is a discount factor

This estimator uses the identity:$\nabla_\theta \mathbb{E}[R] = \mathbb{E}[R \cdot \nabla_\theta \log\pi_\theta]$

$\theta \in \mathbb{R}^p$	Parameter vector of the policy π_θ
R	Reward (or return) — can be a scalar random variable dependent on actions sampled from the policy
π_θ	Probability distribution (policy) from which actions are drawn, given the current state
$\nabla_\theta \mathbb{E}[R]$	Gradient of the expected reward with respect to parameters θ
$\nabla_\theta \log\pi_\theta$	Gradient of the log probability — also called the score function
Term	Meaning

This estimate has high variance → hence noisy

That's why one uses convergence criteria like:

- Smoothing reward signals
- Checking relative changes in reward
- Early stopping if average return plateaus

VI.7 Convergence criteria for

$$\underbrace{max}_{\theta \in \mathbb{R}^p} \mathbb{E}\left[\int_0^T r(t)dt\right]$$

To determine when the algorithm has "converged", you need criteria that ensure the optimization process is close enough to an optimal solution — especially since the model may involve stochasticity, differential equations, and learning dynamics.

VI.7.1 Gradient-Based Convergence (First-Order Optimality)

The most common convergence check for smooth problems:

$$||\nabla_\theta J(\theta)|| \leq \epsilon$$

Where:

- $J(\theta) := \mathbb{E}\left[\int_0^T r(t,\theta)dt\right]$
- ε>0: a small tolerance (e.g.,10^{-5})
- $\nabla_\theta J$: the gradient of the expected reward w.r.t. (with respect to) the parameters θ

This says: stop when the gradient is small, i.e. near a stationary point.

VI.7.2 Relative Improvement Criterion

Check for diminishing returns across iterations:

$$\boxed{\frac{|J^{k+1} - J^k|}{|J^k|} \leq \delta}$$

Where:

- J^k is the objective value at iteration k
- δ: small threshold (e.g., 10^{-5})
- Especially useful when gradient computation is noisy (e.g., using sampling or RL/Reinforcement Learning)

VI.7.3 Paraneter-Change Criterion

Stop if updates to parameters are small:

$$\|\theta^{k+1} - \theta^k\| \leq \eta$$

Where η is a small scalar (e.g., 10^{-7}).
This guards against wasting compute once the parameters are no longer meaningfully evolving.

VI.7.4 Stochastic Methods: Variance in Estimates

In stochastic optimization (e.g., reinforcement learning or Monte Carlo estimation of $\mathbb{E}$: Convergence can also be based on variance stabilization:

$$\boxed{Var\left(\int_0^T r(t,\theta)dt\right) < \varsigma}$$

Symbol	Meaning
ς	Small positive threshold controlling **allowed variance** in reward estimates
Purpose	Ensures convergence check is robust under noisy (stochastic) estimates
Use case	Common in reinforcement learning, stochastic optimization, Monte Carlo methods

Setting	Typical ς
High-precision control	10^{-6} *to* 10^{-4}
RL or stochastic simulation	10^{-3} *to* 10^{-1}
Coarse learning or exploration	10^{-1}

Or:

- Confidence intervals of reward estimates become narrow.
- Reward vs. iteration plot stabilizes.

VI.7.5 Physics-Based Convergence

If this model involves mechanics (e.g., straightening a beam):

- Residual forces at nodes become small
- Strain energy stabilizes or minimizes:

$$||\delta E_{strain}/\delta\theta|| \to 0$$

- Target geometry is reached within tolerance

VI.8 Functional Dependency and Effort Fields

The analysis of functional dependency [see also [1,2,3]] is one of the author´s favorite "mathematical technology" to get close to causality. The use of the entropy theory will make the subject of a later analysis and is not an object of the present work.

- calculation Example (this is merely a mathematical example and has no direct practical relevance since the geometrical tolerances and physical characteristics are not taken into consideration).

Let control points be

$P_1=(1,1,1)$, $P_2=(2,2,1)$, $P_3=(-1,-2,0.5)$

Evaluate at $x=(0,0,0)$

Effort gradient function for each control point Pj:

$$effortGradient(P_j, x, \epsilon) = \nabla f_j(x) = -5\epsilon^2 \frac{\|x - P_j\|^3 (x - P_j)}{\left(\|(x - P_j\|^5 + \varepsilon^2\right)^2}$$

The scalar effort field:

$$f_j(x) = \frac{\epsilon^2}{\left(\left\|(x - P_j\right\|^5 + \varepsilon^2\right)^2}$$

This is a localized field centered at the control point P_j. It:

- Peaks near P_j
- Decays rapidly with distance
- Is smoothed by ϵ to avoid singularities

$\nabla f_j(x)$ It tells that the direction and rate of change of the effort field at point x. The negative gradient vector points toward P_j — that is, toward higher effort.

This makes sense physically: if one applies a force at P_j , other points x will feel a "pull" toward it — that's what the gradient describes.

Term-by-term breakdown

- $r = x - P_j$: the vector from the control point to x
- $\|r\|$=distance from x to P_j
- $\nabla f_j(x) = -5\epsilon^2 \frac{\|r\|^3 r}{(\|r\|^5 + \varepsilon^2)^2}$
- Numerator $\|r\|^3 \cdot r$: This gives direction and scales with distance³, so the field changes rapidly near the source.
- Denominator$(\|r\|^5 + \varepsilon^2)^2$: Makes the whole expression decay smoothly with distance, and ϵ^2 ensures the function doesn't blow up at r=0.
- Negative sign: The gradient points toward the source point P_j, which aligns with how forces tend to "attract" material in physical models.

This particular structure is inspired by mollified (regularized) potential fields often used in:

- Soft body deformation
- Robotics path planning
- Continuum mechanics
- Physics-based animation

It ensures:

- Smoothness everywhere (no singularities)
- Localized influence (decays quickly with distance)
- Differentiability for numerical simulation

Jacobian matrix J: $J = \begin{bmatrix} \nabla f_1(x) \\ \nabla f_2(x) \\ \nabla f_3(x) \\ \nabla \mathrm{h(x)} \end{bmatrix}$ $where\ \nabla \mathrm{h(x)} = \nabla f_1(x) + \nabla f_2(x) + \nabla f_3(x)$

Explicitly, if $\nabla f_j(x) = \left(\frac{\partial f_j}{\partial x}, \frac{\partial f_j}{\partial y}, \frac{\partial f_j}{\partial z}\right)$

$$\text{Then } J = \begin{bmatrix} \partial_x f_1 & \partial_y f_1 & \partial_z f_1 \\ \partial_x f_2 & \partial_y f_2 & \partial_z f_2 \\ \partial_x f_3 & \partial_y f_3 & \partial_z f_3 \\ \sum_{j=1}^{3} \partial_x f_j & \sum_{j=1}^{3} \partial_y f_j & \sum_{j=1}^{3} \partial_z f_j \end{bmatrix}$$

Summary:

- Rows 1-3: gradients of individual effort functions from each control point at the evaluation position.
- Row 4: gradient of total effort $h=\sum f_j$
- This 4-by-3 matrix is then analyzed for rank and eigenvalues of $J^T J$.

Jacobian Matrix (Rows = ∇f_1, ∇f_2, ∇f_3, ∇h)	$\begin{pmatrix} 0.0010678 & 0.0010678 & 0.0010678 \\ 0.000045721 & 0.000045721 & 0.0000228605 \\ -0.000150756 & -0.000301512 & 0.0000753781 \\ 0.000962762 & 0.000812005 & 0.00116604 \end{pmatrix}$
Rank of Jacobian (rank_J)	3
Eigenvalues of $J^T J$ (λ_1, λ_2, λ_3)	$\{6.37664 \times 10^{-6}, 1.13776 \times 10^{-7}, 7.46648 \times 10^{-11}\}$

What the code does:

- Defines control points $P_j \in R^3$ and evaluation point x.
- Computes effort gradients $\nabla f_j(x)$ at that point from each control point P_j, incorporating a smoothing parameter ϵ to avoid singularities.
- Constructs the Jacobian matrix J, where each row is $\nabla f_j(x)$ and the last row is their sum $\nabla \mathrm{h(x)} = \nabla f_1(x) + \nabla f_2(x) + \nabla f_3(x)$ (in the case of 3 components)
- Calculates the rank of J and eigenvalues of $J^T J$
- Outputs formatted results for interpretation.

VI.9 Gradient Field

- Jacobian matrix J encodes local sensitivity of the displacement field at node x due to forces at control points.
- Rank of J:
  - Full rank (=3 in 3D) means control points independently affect all directions at x — no convergence yet.
  - Rank < 3 means dependent directions or locked (Degrees of Freedom)DOFs — node is close to equilibrium.
- Eigenvalues of $J^T J$
- Small eigenvalues (close to zero) imply directions with little or no response — **signal convergence.**
- Large eigenvalues indicate directions still actively influenced

How to use this in the straightening algorithm:

1. At each node x_i, run this computation.
2. If Rank(J)<3 or min(eigvals)<δ, mark node i as converged.
3. Adjust forces or impulses applied accordingly to focus effort on unconverged nodes.
4. Iterate until all nodes are converged.

Possible next steps:

- Extend this code to automate checking multiple nodes along the piece.
- Couple this with time-stepping and force update loops to simulate the full straightening.
- Use these convergence tests to adaptively refine control point force magnitudes $\alpha_j(t)$.

Fig.1:

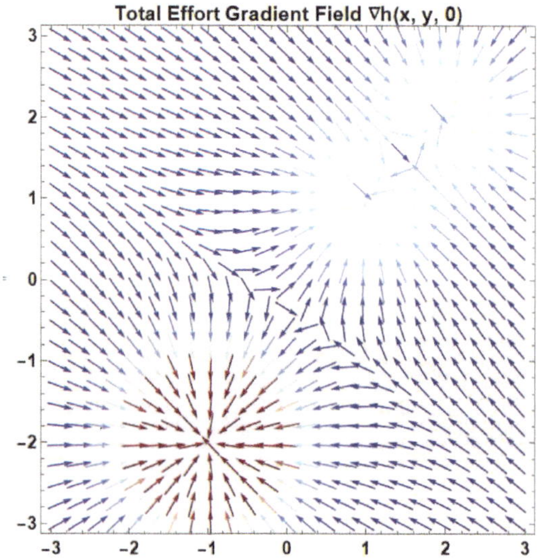

The sum of effort gradients coming from multiple control points $\{P_1, P_2, P_3\}$ evaluated over the 2D plane z = 0 (see fig. 1). This is a vector field — at each point (x, y) in the domain, the arrow shows:

- Direction: where the effort "pulls" the material point
- Magnitude: how strong that pull is

The color of each arrow reflects the magnitude of the total gradient at that point

Physical Interpretation for Straightening (see fig. 2 and 3)

In the given context (mechanical straightening of a deformable body), this plot (see fig. 2) tells:

✅ 1. Direction of Movement

- Each point in the domain "feels" a net influence from all control points.
- The direction of the arrow shows where the material would tend to move or deform under current force settings.

✅ 2. Zones of Influence

- Regions closer to a control point show stronger gradients (brighter colors, longer arrows).
- Where control point influences overlap, interference occurs (arrows bend or cancel).

✅ 3. Gradient Flow = Straightening Flow

- The material will deform along the gradient flow lines — following the arrows.
- By following the field lines, you could define smart straightening paths.

✅ 4. Critical Points (Small Gradients)

- Areas with very small vectors (cooler colors or nearly zero length) indicate:
  - Equilibrium zones
  - Possibly already straightened or balanced areas
  - Or regions where forces cancel out

Feature	Meaning
Arrow direction	Pulling direction due to all P_1, P_2, P_3
Arrow length / color	Force magnitude: how strongly the point is influenced
Converging arrows	Node is being pulled toward a local straightening center
Diverging arrows	Node is being pushed away — high curvature or stress region

Fig.2:

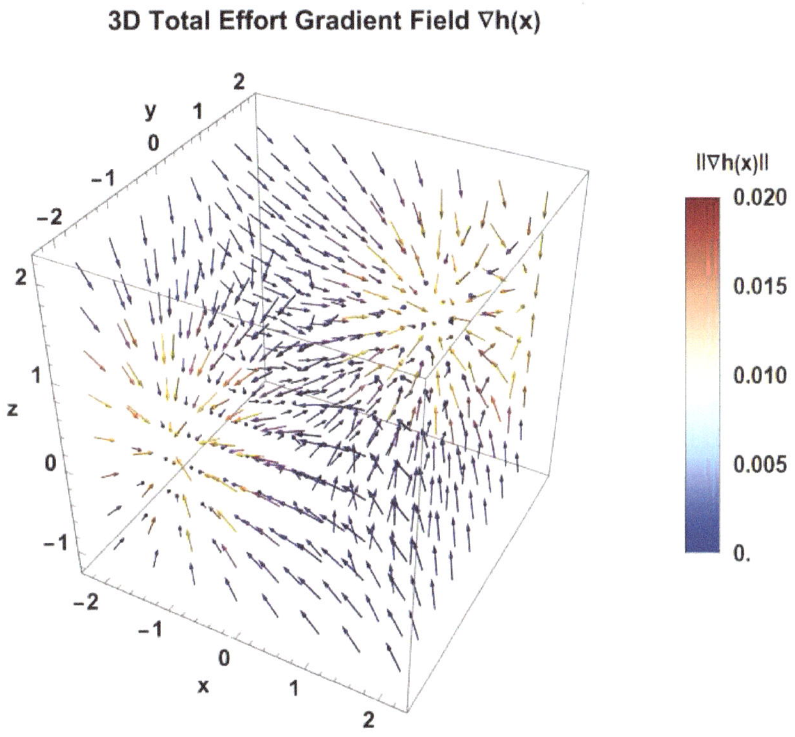

What This Tells About the Straightening Strategy

- Gradient Vectors (∇h) show the direction and magnitude of "pull" due to all control points.
- Where the field is strong (bright colors), the material/node is likely far from equilibrium and still being influenced.
- Weaker regions (faded arrows) indicate converging or already-straightened zones.
- You can identify zones of influence from each control point and assess how efficiently nodes are straightening based on the direction and density of gradient lines.
- This can help adapt your control strategy, for instance, applying additional force in under-affected regions.

Fig.3: direction and magnitude of the total effort gradient at grid points.

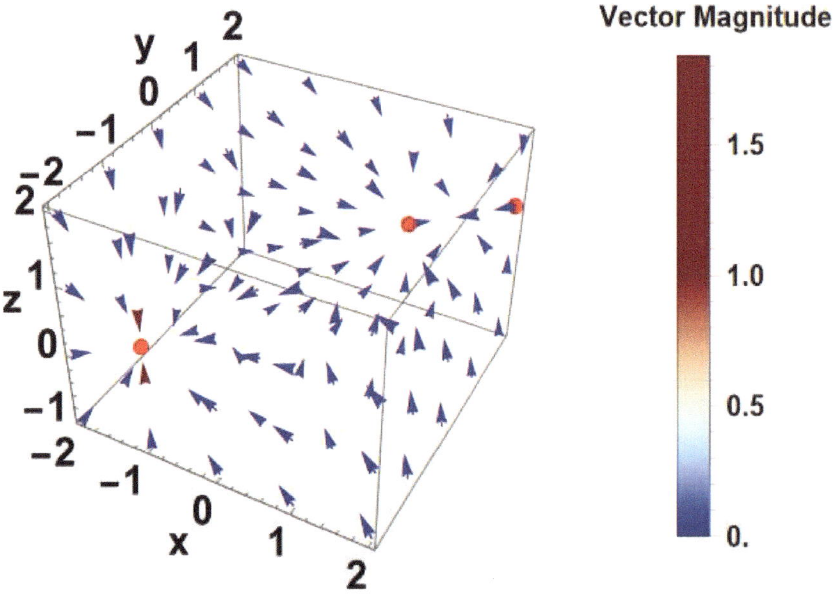

- The 3D arrows represent the direction and magnitude of the total effort gradient at grid points.
- Red points are the control points.
- The legend bar shows the color-to-magnitude mapping.

Physical Meaning in Straightening

Each arrow in the vector field represents the force direction and relative intensity acting on a point in space due to the superposition of all control points' influence.

Interpretation:

The vector at position x is $\nabla h(x)$ — the gradient of the effort field.

The effort field is a proxy for how much "energy" or "deformation pressure" is experienced at a point.

The gradient (∇h) is the direction of steepest increase of effort, so it acts like a corrective force pushing the body back to equilibrium.

Strategy Implications

1. Direction of Correction

- The vectors point in the direction the material/node should move to reduce effort (i.e., to straighten).
- Applying force against the gradient is equivalent to pushing the node toward mechanical balance.

2. Magnitude = Urgency

- Long arrows (large gradient magnitude) indicate strong imbalance — areas that are far from convergence or equilibrium.
- Short arrows suggest the node is already well-aligned or close to equilibrium.

3. Convergence Detection

- If the gradient vector is nearly zero, the node is in a stable or balanced position.

This is used in your algorithm's convergence check: a near-zero gradient implies local mechanical equilibrium.

Self-Adjusting Force Strategy

Since you can visualize where each part of the body wants to go (via ∇h), you can:

- Adaptively scale applied impulses: stronger where gradient is large.
- Choose where to apply new control impulses (P_j) based on regions with high remaining gradient.
- Iteratively update control point weights $\alpha_j(t)$ to steer the body more efficiently.

Learning Component

Once have this physical model ready:

- A learning algorithm (e.g. reinforcement learning) can:
    - Observe gradient behavior,
    - Learn which control point configurations lead to fast convergence,
    - Optimize the control strategy (selection, timing, force amplitude).

Gradient magnitude field — it's one of the most insightful scalar quantities you can compute from the total effort gradient ∇h(x).

$$\nabla h(x) = \sum_{j=1}^{k} \nabla f_i\,(x)$$

$$||\nabla h(x)|| = \sqrt{\left(\frac{\partial h}{\partial x}\right)^2 + \left(\frac{\partial h}{\partial y}\right)^2 + \left(\frac{\partial h}{\partial z}\right)^2}$$

This tells that the strength of the total "pulling force" at each point in space — nodes will move fastest where ||∇h(x)|| is large.

Fig. 4:

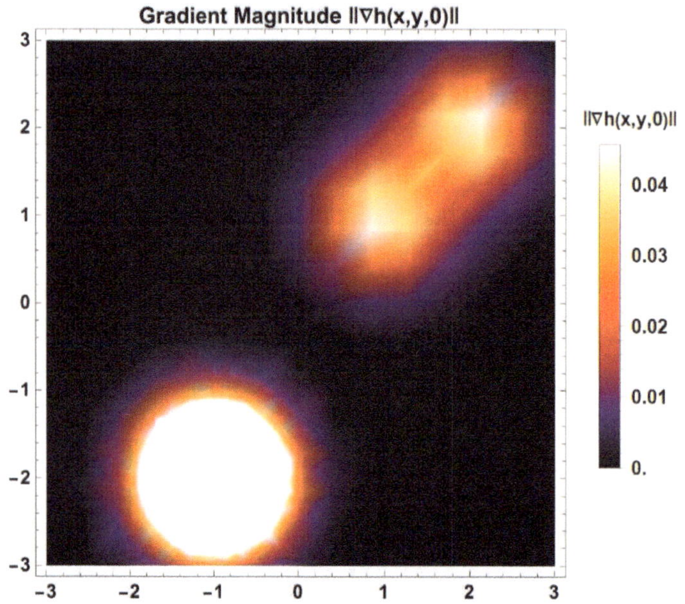

Interpretation Straightening Strategy

- Low ‖∇h(x)‖ Indicates areas where nodes are already close to equilibrium.
    - ⟶ Good convergence regions
    - ⟶ These nodes can be marked as converged
- High ‖∇h(x)‖ Indicates strong residual force
    - ⟶ Still needs correction
    - ⟶ High-risk for instability if not damped

Fig. 5:

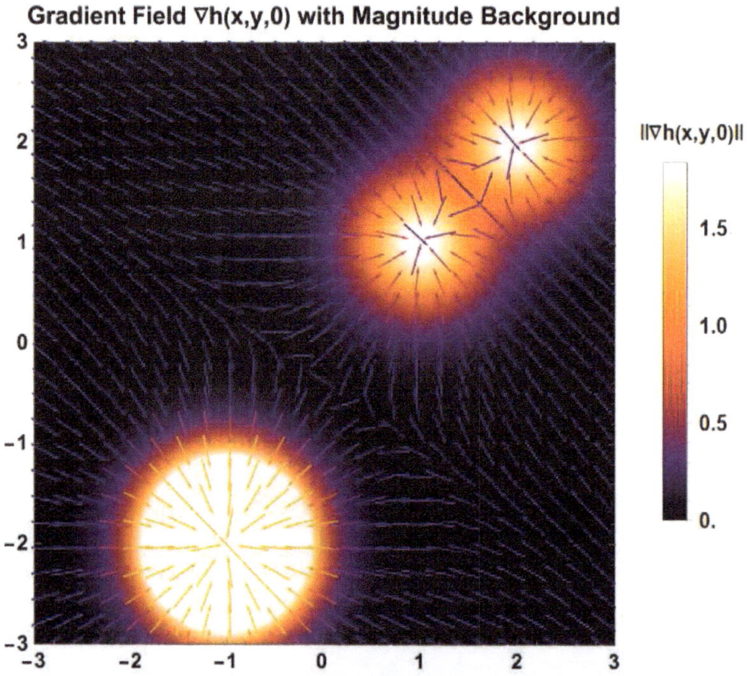

Interpretation:

- The second algorithm (fig. 5) is more informative, because it visualizes the vector field ∇h (i.e. direction of learning or force) together with its magnitude as background color.
- The first algorithm (fig. 4) is less detailed, giving only a heatmap of how "steep" the effort landscape is, without directionality.
    - Red/white zones = large $\|\nabla h\|$: the node feels a strong effort force.
    - Dark/blue zones = near-zero $\|\nabla h\|$: candidate for **convergence**.
- Arrows:
    - Indicate the direction in which the node would move.
    - Can guide selection of where to apply correction impulses.

Fig. 6:

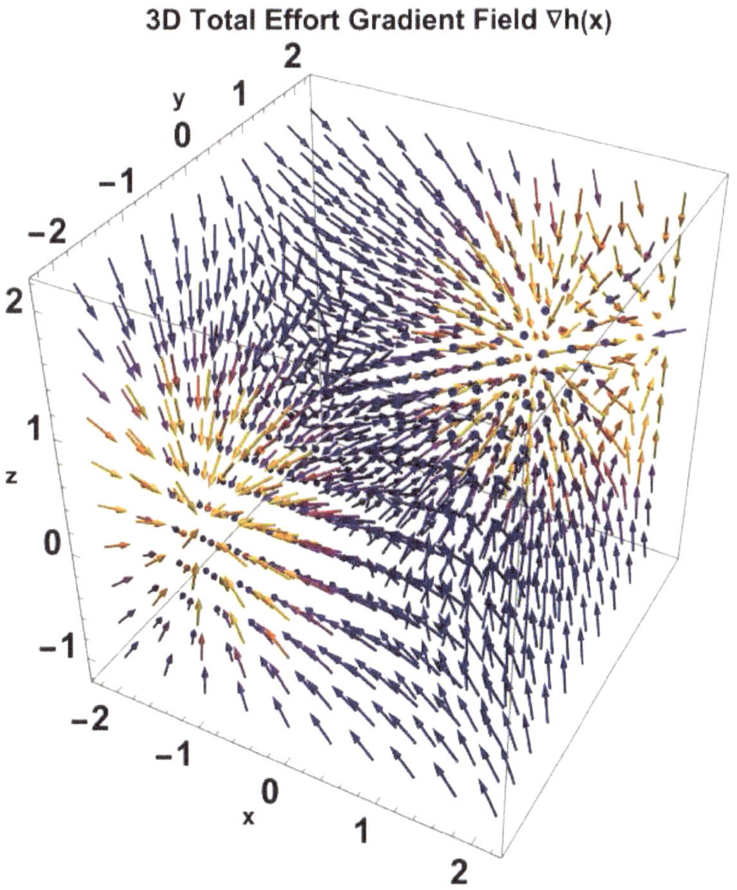

Fig. 7:

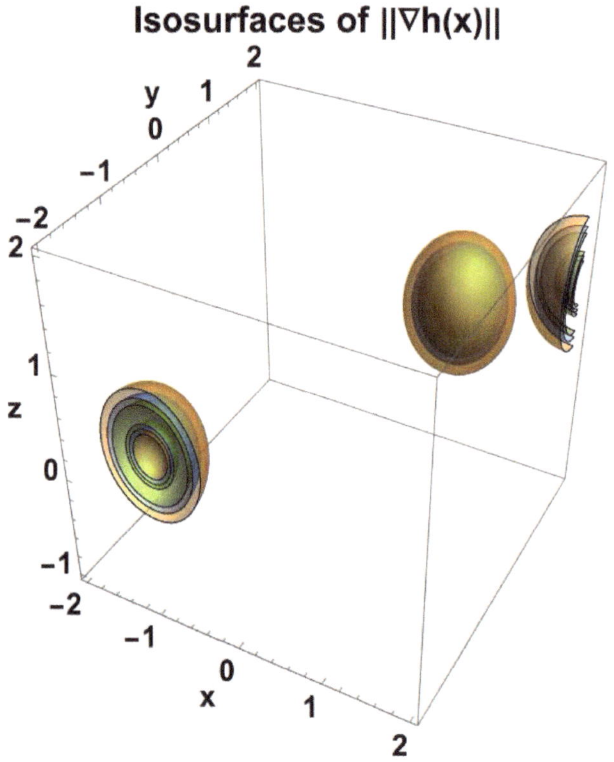

Interpretation for Straightening Strategy

- Streamlines (fig. 6): Show how each node or part of the body would flow or move if following the force field.
    - Curved, converging streams → coordinated straightening.
- Isosurfaces (fig. 7): Regions of equal effort magnitude.
    - Can guide adaptive control: apply stronger impulse in high-gradient zones.

VI.10 Markov Decision Process

Example of calculation – pure mathematical formulation of a self-learning straightening model using a Markov Decision Process (MDP) and Q-learning, including a convergence criterion.

For a body having three points which has to be straightened.

Definitions

VI.10.1 Setup

- State space:

$$S = \{s \in \{0,1\}^3\}$$

Each state $s=(s_1,s_2,s_3)$ where $s_i \in \{0,1\}$

$s_i=0$: point i is aligned

$s_i=$: point i is misaligned

So: |S|=8

Action space: $A = \left\{a_J : J \subseteq \{1,2,3, J \neq \emptyset\}\right\}$

Each action a_J hits a subset J of the 3 points. So |A|=7.

Transition Probability:

P(s'|s,a)=Pr(next state is s'|current state s,action a)

This is stochastic and depends on the physical process.
Example: $P(000|101,a_{\{1,3\}})=0.8$

Reward function: $R(s,a,s`) = \begin{cases} +R_{goal} \; if \; s' = (0,0,0) \\ -c \quad otherweise \end{cases}$

Where:

- $R_{goal} \gg 0$ (e.g. 100)
- c>0: cost per action (e.g. 1)

VI.10.2 Learning Algorithm: Q-Learning

We define a value function $Q: S \times A \to \mathbb{R}$

Q-Update Rule:

At iteration t:

Let current state be s_t, action a_t, next state s_{t+1}, reward r_t

$$Q_{t+1}(s_t,a_t) = Q_t(s_t,a_t) + \alpha \left[r_t + \gamma \underbrace{max}_{a'} Q_t(s_{t+1},a') - Q_t(s_t,a_t) \right]$$

Where:

- $\alpha \in (0,1)$: learning rate
- $\gamma \in [0,1)$: discount factor
- $a' \in A$: all possible next actions

  Policy

At any time t, the greedy policy is:

$$\pi_t(s) = arg \underbrace{max}_{a \in A} Q_t(s,a)$$

An exploration strategy (like ε-greedy) is used:

$$\pi_t^{\epsilon}(s) = \begin{cases} random \; a \in A & with \; probability \; \epsilon \\ argmax_a Q_t(s,a) & with \; probability \; 1-\epsilon \end{cases}$$

VI.10.3 Convergence Check

Let: $\Delta_t = \underbrace{max}_{s \in S, a \in A} |Q_{t+1}(s,a) - Q_t(s,a)|$

Then define convergence when: $\Delta t < \delta$ for some small threshold $\delta > 0$ (e.g. $\delta = 10^{-3}$).

Termination Condition for Straightening Process

The process is considered successfully completed (converged in application) when: s=(0,0,0)

That is, all three points are aligned.

Summary

solving a finite MDP:

- State space: $S=\{0,1\}^3$
- Action space: A={hit any nonempty subset of {1,2,3}}
- Learn Q(s,a) using the update rule
- Apply the greedy policy
- Stop when convergence criterion met:
  - For learning: $\Delta t < \delta$
  - For process: s=(0,0,0)

VI.10.4 Matrix-Based Formulation of the MDP

1. State Space S

States are all 3-bit binary combinations:

Index State	(Bit vector)
0	(0,0,0)
1	(0,0,1)
2	(0,1,0)
3	(0,1,1)
4	(1,0,0)
5	(1,0,1)
6	(1,1,0)
7	(1,1,1)

Define a canonical order: $S = \{s_0, s_1, s_2, s_3, s_4, s_5, s_6, s_7\} \in \mathbb{R}^8$

2. Action Space A

Action a_j	Description
a_1	Hit point 1
a_2	Hit point 2
a_3	Hit point 3
a_{12}	Hit points 1 and 2
a_{13}	Hit points 1 and 3
a_{23}	Hit points 2 and 3
a_{123}	Hit all 3 points

So: $A=\{a_1,a_2,\ldots,a_{123}\}$, cardinality 7.

3. Transition Matrices P^a

For each action a∈A, define an 8×8 transition matrix:

$$P^a = [p_{ij}^a] \in \mathbb{R}^{8x8}, where\ p_{ij}^a = P_r(s_j \mid s_i, a)$$

Each row sums to 1: $\sum_{j=0}^{7} p_{ij}^a = 1\ \forall i$

Notation	Meaning
$P_r(s_j \mid s_i, a)$	Probability of transitioning to s_j from s_i using action a
p_{ij}^a	The same as above, entry (i,j) in matrix P^a
$P^a \in \mathbb{R}^{8x8}$	Transition matrix for action a
$\sum_j P_{ij}^a = 1$	Rows of P^a sum to 1 (valid probability distribution)

Example for Action a_{13} (hit points 1 and 3):

Let's say the hitting effect is probabilistic:

- A misaligned point becomes aligned with 90% probability
- An aligned point stays aligned with 95% probability
- Probability of overcorrection (aligned → misaligned): 5%
- Probability of no effect: remaining mass

Define these for point transitions and then derive the full matrix.

Example:

$s_i=(1,0,1) \Rightarrow s_5$

Apply a_{13} : hitting points 1 and 3

- Point 1: 1 → 0 with 0.9
- Point 3: 1 → 0 with 0.9

Assuming independence (here is an important point of the strategy, if not independent, then strategy adaptation!):

- Probability of reaching (0,0,0) = 0.9×0.9=0.81
- Probability of point 3 stays 1: 0.1
- So (0,0,1): 0.9×0.1=0.09
- (1,0,0): 0.1×0.9=0.09
- (1,0,1): 0.1×0.1=0.01

So: $P^{a_{13}}[5,:] = [0.81,0.09,0,0,0.09,0.01,0,0]$

Build similar rows for all $s_i \in S$ to complete the matrix.

4. Reward Matrix R

Define reward for each (s,a,s') triple:

$$R(s_i, a_i, s_k) = \begin{cases} +R_g & if\ s_k = s_0 = (0,0,0) \\ -c & otherweise \end{cases}$$

Alternatively, one can define a reward vector $r^a \in \mathbb{R}^8$ for each action:

$$r_i^a = \mathbb{E}R[(s_i, a, s')] = \sum_{j=0}^{7} P^a[i,j] \cdot R(s_i, a, s_j)$$

5. Value Iteration (Matrix Form)
   Let $V \in \mathbb{R}^8$: state value function
   Value Iteration Update:

$$V^{t+1} = \underbrace{max}_{a \in A} \{r^a + \gamma P^a V^t\}$$

   This uses:

- r^a: expected immediate reward for action a
- γ: discount factor
- $P^a V$: expected future value under action a

   Convergence when:

$$\|V^{t+1} - V^t\|_\infty < \delta$$

6. Q-Function (Matrix Form)

$$Q^a = R^a + \gamma P^a V$$

   and policy:

$$\pi(s) = arg\, \underbrace{max}_{a}\, Q^a(s)$$

   Final Notes

- You can construct 7 transition matrices $P^a \in \mathbb{R}^8$
- Each matrix models how state probabilities shift under that action
- Reward vectors $r^a \in \mathbb{R}^8$
- Use value iteration or Q-learning for policy learning
- Converge when policy/value change is below a threshold

VII. Collapsing model

the model collapses all points to their centroid — essentially crushing the "body" into a single point; in a physical sense, the *body would be destroyed* by this straightening because all points merge.

1D self-correcting system of three physical points using principles like:

This is a *form* of Q-learning-style dynamic control using geometry as the reward cue. The model is a deterministic, geometrically-driven control algorithm that:

- Behaves like a hand-crafted RL policy
- Can be generalized into a Q-learning framework
- Already shows adaptive correction based on feedback

Q-learning is a model-free **reinforcement learning (RL)** algorithm that learns the value Q(s,a) of taking an action a in a state s, using the update rule:

$$Q(s_t, a_t) \leftarrow Q(s_t, a_t) + \alpha \left[r_{t+\gamma} + \gamma \max_a Q(s_{t+1}, a) - Q(s_t, a_t) \right]$$

Where:

- Q(s,a): expected future reward of action a in state s
- r_t: reward at time t
- α: learning rate
- γ: discount factor

Is the current Model Q-learning?

Not strictly, but your model shares many structural features with Q-learning — it's a geometric, deterministic analog of a value-based self-learning system.

Here's a comparison:

Concept in Q-learning	Analog in the current Model
State s	Point position p^k
Action a	Hit sets S^k (which points to move)
Reward r	Reduction in deviation ΔD^k
Policy	Hit strategy: chooses hit set based on deviation
Value Function Q(s,a)	Implicit in the rule: which actions lead to fastest convergence
Learning Rate α	Geometric $\alpha^k = min\left(1, \gamma \cdot \frac{D^k}{D^0}\right)$
Convergence criterion	Deviation $D^k < \epsilon$

What's Missing (vs full Q-learning)?

- No stochasticity (everything is deterministic).
- No exploration strategy (like epsilon-greedy).

- No explicit Q(s,a) table or value estimation — the reward function is hard-coded geometrically.

Mathematically Sound, Physically Misguided

The algorithm is:

- Mathematically valid: It reduces deviation from the centroid using a well-defined iterative method. The learning rate, update rule, and convergence criterion are coherent and deterministic.
- Geometrically smooth: It ensures monotonic decrease in the scalar deviation function D^k, guaranteeing convergence (in theory) to a unique equilibrium point.
- Algorithmically elegant: As a control policy, it's clear, efficient, and rule-driven.

But —

- From a physical modeling perspective, it fails to respect the structure or constraints of real-world bodies (e.g. rods, chains, limbs, particles bound by spring forces).
- It leads to complete collapse, which in almost any physical system would mean breakdown or loss of structure.

Why This Is Still Useful

Even though it's not physically sensible *as is*, this model has strong *conceptual value*:

- It illustrates how geometric feedback can drive convergence.
- It shows how a deterministic system can mimic learning behavior without an actual Q-table.
- It could serve as a building block or intuition for better physically-plausible systems.

1. Geometric Centroid
Each iteration pulls selected points toward the centroid of all current positions:

$$c^k = \frac{1}{3}\sum_{i=1}^{3} p_i^k$$

This is the average position. The idea is to move some points toward this average to achieve symmetry.
Let the positions of the three points at iteration k be:

$$p^k = (p_1^k, p_2^k, p_3^k)$$

2. Deviation

$$D^k = \sum_{i=1}^{3} (p_i^k - c_i^k)^2$$

This scalar value measures how "off-center" the configuration is. If$D^k = 0$, all points are at the correct position (perfect straightening).
This is a measure of how “unsymmetric” the system is, and your convergence condition is:

$$D^k < \epsilon$$

3. Learning Rate
Define the learning rate α^k based on how far we are from equilibrium /where $\gamma \in (0,1]$ is a tunable scaling parameter/:

$$\alpha^k = min\left(1, \gamma \cdot \frac{D^k}{D^0}\right)$$

This ensures large corrections when far from convergence, and smooth updates near convergence.

4.Hit Strategy (Dynamic Control)

Based on D^k dynamically select a hit set $S^k \subseteq \{1,2,3\}$ of indices to move:

$$S^k = \begin{cases} \{1,2,3\}, & D^k > 1 \\ top\ 2\ farthest\ from\ c^k, & 0.1 < D^k \leq 1 \\ farthest\ point, & D^k \leq 0.1 \end{cases}$$

This mimics targeted correction: only the most problematic points are moved as the system approaches symmetry.

A smart rule that changes which points to move, depending on how far the system is from equilibrium:

- Large deviation: hit all points
- Moderate deviation: hit 2 farthest
- Small deviation: hit 1 farthest
- Update Rule
- Update the positions of the selected hit set:

$$p_i^{k+1} = \begin{cases} p_i^k + \alpha^k (c^k - p_i^k), & i \in S^k \\ p_i^k, & i \notin S^k \end{cases}$$

This pulls hit points toward the centroid using a controlled step size.

Repeat Until Convergence

Why Convergence After 20 Iterations? (see fig. 1)

1. Initial Deviation is high → the algorithm starts with large moves.
2. As deviation shrinks, the learning rate and number of hit points are reduced.
3. Eventually, only one point is moved and with very small steps, requiring many fine-tuning steps.

Convergence Criterion: The loop stops when $D^k < \epsilon = 0.0001$

This is expected: convergence slows as the points cluster closely and corrections become tiny.

Fig. 1:

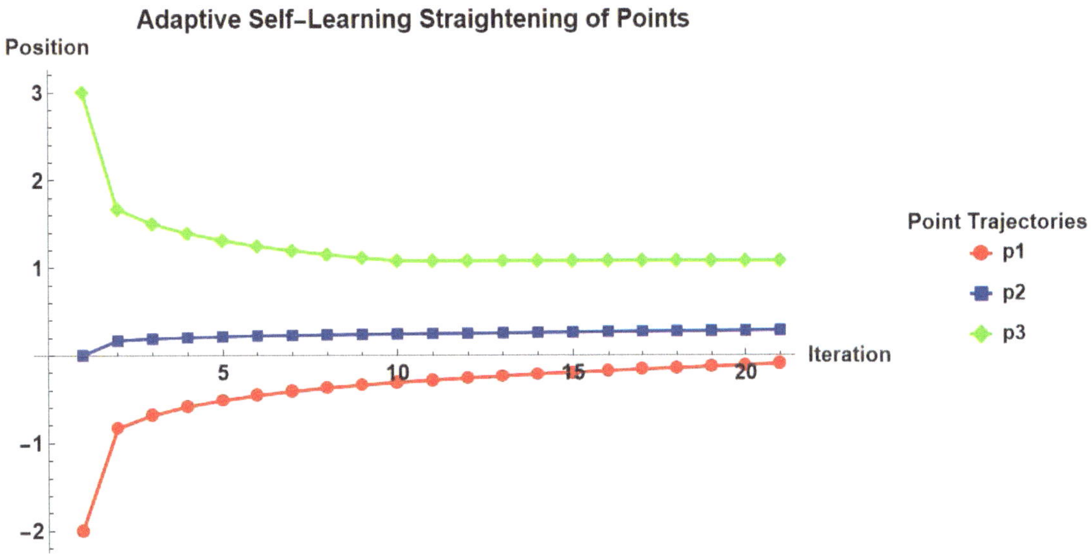

Graphic Interpretation

ListLinePlot correctly shows how:

- All 3 points converge toward the centroid.
- At first, lines move sharply.
- Then, only one point is adjusted (likely the farthest one), leading to small tweaks.
  The lines flatten as convergence is approached — a hallmark of exponential-like decay in deviation.

Iteration log table for your adaptive self-learning straightening process applied to the initial positions { -2, 0, 3 }, using:

- γ=0.5
- ϵ=0.001
- $max_{Iteration}$=20
- Converged after 20 iterations with deviation 0.717590
- Final positions: {-0.100050, 0.282750, 1.074400}

Iteration	Positions	Hit Set	Deviation	Learnig rate
0	{-2.000, 0.000, 3.000}	{1, 2, 3}	7.555556	1.0000
1	{-0.667, 0.111, 1.000}	{1, 2, 3}	1.192901	0.0789
2	{-0.148, 0.111, 0.630}	{1, 3}	0.157206	0.0104
3	{-0.120, 0.111, 0.418}	{1, 3}	0.072066	0.0048
4	{-0.104, 0.111, 0.314}	{1, 3}	0.038979	0.0026
5	{-0.093, 0.111, 0.259}	{1, 3}	0.024067	0.0016
6	{-0.085, 0.111, 0.226}	{1, 3}	0.017363	0.0011
7	{-0.078, 0.111, 0.206}	{1, 3}	0.013824	0.0009
8	{-0.073, 0.111, 0.193}	{1, 3}	0.011950	0.0008
9	{-0.069, 0.111, 0.185}	{1, 3}	0.010854	0.0007
10	{-0.066, 0.111, 0.180}	{1, 3}	0.010073	0.0007
11	{-0.064, 0.111, 0.176}	{1, 3}	0.009470	0.0006
12	{-0.063, 0.111, 0.174}	{1, 3}	0.008990	0.0006
13	{-0.061, 0.111, 0.172}	{1, 3}	0.008596	0.0006
14	{-0.060, 0.111, 0.171}	{1, 3}	0.008266	0.0005
15	{-0.059, 0.111, 0.170}	{1, 3}	0.007983	0.0005
16	{-0.059, 0.111, 0.170}	{1, 3}	0.007737	0.0005
17	{-0.058, 0.111, 0.169}	{1, 3}	0.007518	0.0005
18	{-0.058, 0.111, 0.169}	{1, 3}	0.007320	0.0005
19	{-0.057, 0.111, 0.169}	{1, 3}	0.007137	0.0005
20	{-0.057, 0.111, 0.169}	{}	0.000999	—

The following picture (fig. 2) shows a 3D collapsing algorithm; all 3 points converge toward the centroid.

Fig.2:

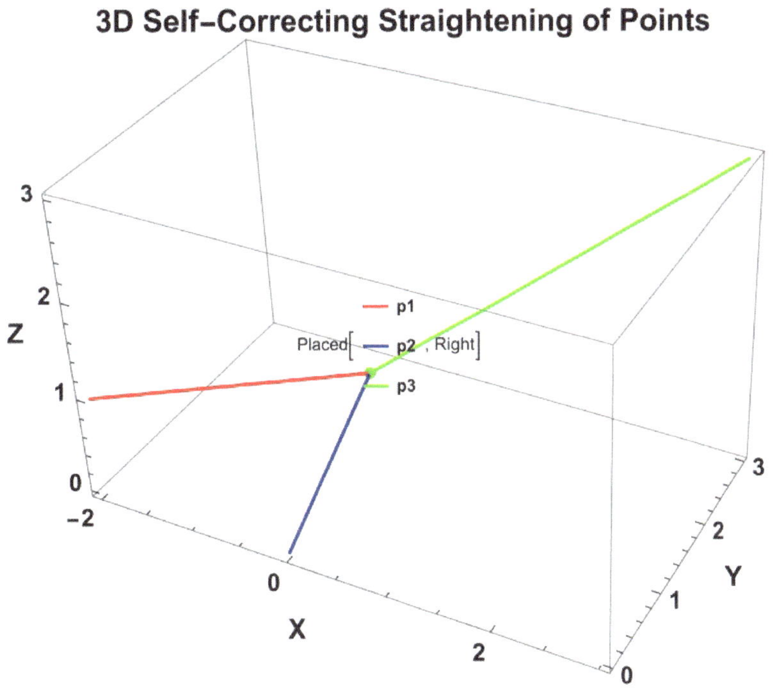

Converged after 100 iterations with deviation 3.11029000
{{-2, 0, 1}, {0, 0, 0}, {3, 3, 3}}
Final positions:
{{0.333122, 0.999910, 1.333300}, {0.333303, 0.999910, 1.333210}, {0.333574, 1.000180, 1.333480}}

Does this collapse of a 3D object to a point bear an abstract or conceptual similarity to the Poincaré Conjecture, which asserts that a simply connected 3D manifold without boundary is equivalent (homeomorphic) to a 3-sphere — and topologically collapsible to a point?

Yes — abstractly and metaphorically, the algorithm's collapse of a 3D structure to a point mirrors the intuition behind the Poincaré Conjecture.

But not in the mathematical/topological sense, because:

- *the algorithm is geometric and discrete, not continuous or topological.*
- *Poincaré deals with continuous deformation (homotopy), not destructive convergence.*
- *collapse destroys structure; Poincaré's deformation preserves topological identity.*

Let's Compare More Precisely

3D Body

- *Made of 3 points in 3D space.*
- *Initially forms a simple 3D structure (like a spatial triangle or a tetrahedral corner).*
- *apply an adaptive, deterministic geometric rule (based on centroid or PCA) to straighten or symmetrize.*
- *The algorithm iteratively moves the points.*
- *Eventually, all points collapse into a single point (or get arbitrarily close).*
- *The original spatial extent — the "body" — is reduced to nothing.*

Poincaré Conjecture

- *A statement about 3D manifolds (smooth, connected, without boundary).*
- *Says: If a 3D manifold is simply connected (every loop can be contracted to a point), then it's topologically a 3-sphere.*
- *A 3-sphere is contractible to a point in a topological sense.*
- *Perelman's proof shows this via Ricci flow — which smoothly deforms the shape until it becomes a round sphere (and topologically a point).*

So, What's the Abstract Connection?

Concept	**3D Collapse**	**Poincaré Conjecture**
3D object	Yes — 3 non-colinear points forming a simple spatial structure	Yes — 3-manifold
No boundary	Implicit — points are free in space	Explicit requirement
No holes (simple connectivity)	Yes — trivially true for 3 points	Required
Convergence to a point	Yes — points literally collapse	Yes — sphere is topologically contractible
Process type	Discrete, geometric, destructive	Continuous, topological, structure-preserving
Final state	Single point (0D)	Topological point (via 3-sphere)
Interpretation	Geometric degeneracy	Topological equivalence

Key Differences

- *The algorithm is not about topology; it doesn't track connectivity, loops, or holes.*
- *Collapse in the model is numerical and mechanical, not topological.*
- *In Poincaré, collapsing to a point preserves structure in the limit; in your algorithm, structure is destroyed by merging points*

Philosophical Interpretation

In both cases, we see a system where a seemingly complex 3D form — whether it's a manifold or a cloud of points — is shown, through internal rules, to be reducible to something trivial: a point.

That "everything collapses to unity" intuition is at the heart of both — one in pure topology, one in discrete geometry. From form to formlessness, from geometry to singularity.

VIII. Avoiding collapse

Straighten 3 points in 3D space, making them colinear and symmetrically arranged about their centroid — without collapsing all points to a single location.

Key Concepts

1. Point Positions

Let the points be: $p_1^k, p_2^k, p_3^k \in \mathbb{R}^3$
at iteration k, where: $p^k = [p_1^k, p_2^k, p_3^k]$

2. Centroid

$$c^k = \frac{1}{3}\sum_{i=1}^{3} p_i^k$$

3. Deviation (from straightness)

Instead of using distance to centroid (which leads to collapse), use deviation from colinearity:

$$\boldsymbol{D^k} = \|(\boldsymbol{p_3^k} - \boldsymbol{p_2^k}) \times (\boldsymbol{p_1^k} - \boldsymbol{p_2^k})\|$$

- This is the squared area of the triangle formed by the 3 points.
- If the points are colinear, this deviation is zero.
- This avoids collapsing all points to the same spot.

4.Adaptive Learning Rate

$$\alpha^k = min\left(1, \gamma \cdot \frac{D^k}{D^0}\right), \gamma \in (0.1]$$

This ensures fast movement early on, and fine-tuning as the system converges.

5. Hit Strategy

Use the area-based deviation $\boldsymbol{D^k}$ to determine how aggressively to move the points.

$$\boldsymbol{S^k} = \begin{cases} \{\boldsymbol{1, 2, 3}\}, & \boldsymbol{D^k > 1} \\ \boldsymbol{top\ 2\ farthest\ from\ c^k}, & \boldsymbol{0.1 < D^k \leq 1} \\ \boldsymbol{farthest\ point}, & \boldsymbol{D^k \leq 0.1} \end{cases}$$

The "best-fit line" goes through the current centroid and direction vector

For simplicity, the direction can be approximated by $d^k = p_3^k - p_1^k$

6. Projection to Line (Target)

For each selected point $i \in \boldsymbol{S^k}$ compute the projection onto the current line:
Let:

- Line origin: $\boldsymbol{c^k}$
- Line direction: $d^k = p_3^k - p_1^k$

Project: $t_i^k = c^k + \left[\frac{(p_i^k - c^k)}{\|d^k\|^2}\right] d^k$

7. Update Rule

Move only the selected points $i \in \boldsymbol{S}^k$ toward their target projections:

$$p_i^{k+1} = \begin{cases} p_i^k + \alpha^k(t_i^k - p_i^k), & i \in \boldsymbol{S}^k \\ p_i^k, & i \notin \boldsymbol{S}^k \end{cases}$$

8. Convergence Criterion

$$\boldsymbol{D}^k < \boldsymbol{\epsilon}$$

Stops when the triangle's area is small enough to call the points colinear.

Full Algorithm Loop

Input:

- Initial points p_1^0, p_2^0, p_3^0
- Parameters: $\gamma \in (0.1], \boldsymbol{\epsilon} \ll \mathbf{1}, \boldsymbol{max\ iterrations\ N}$
- Compute centroid:

$$c^k = \frac{1}{3}\sum_{i=1}^{3} p_i^k$$

- Compute deviation:

$$\mathrm{D}^k = \|(\mathrm{p}_3^k - \mathrm{p}_2^k) \times (\mathrm{p}_1^k - \mathrm{p}_2^k)\|$$

- Compute learning rate:

$$\alpha^k = min\left(1, \gamma \cdot \frac{D^k}{D^0}\right), \gamma \in (0.1]$$

- Determine direction:

$$d^k = p_3^k - p_1^k$$

- For each i, compute projection t_i^k
- Determine hit set $\boldsymbol{S}^k$ based on D^k
- Update positions of $i \in S^k$
- Check convergence: if $D^k < \epsilon$ stop
- k=k+1

Optional Enhancements

- Log deviation and positions each iteration.
- Visualize in 3D space to confirm the straightening effect.
- Use PCA (Principal Component Analysis) for better axis estimation than $p_3 - p_1$

Key Advantages

- Physically intuitive — maintains spatial spread, avoids collapse.
- Geometrically rigorous — uses actual shape deviation (triangle area).

- Deterministic and adaptive — mimics learning behavior without full Q-table.

Sample Output (first few iterations) (see also fig. 1,2 and 3):

Iteration	Positions	Hit Set	Deviation	Learning Rate
0	{{-2,0,1}, {0,0,0}, {3,3,3}}	{1,2,3}	13.5	1.000
1	{{-1.274,0.274,1.274}, {0,0,0}, {2.274,2.274,2.274}}	{1,2,3}	1.9772	0.0732
2	{{-1.224,0.246,1.196}, {0,0,0}, {2.181,2.225,2.225}}			

Fig.1:

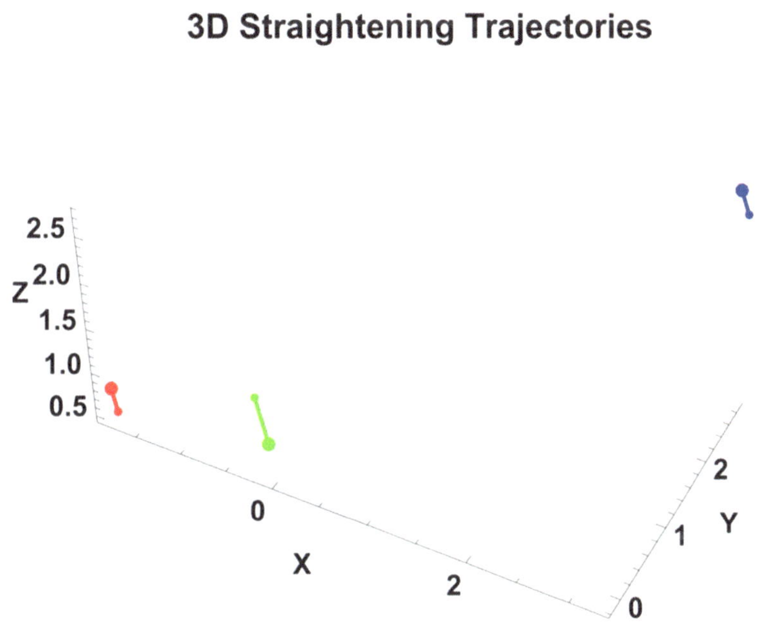

3D plot showing the paths of the 3 points from their initial to final positions.
Red, Green, Blue lines for Points 1, 2, and 3.
Large dots: starting positions.
Medium dots: converged (final) positions.

Fig.2: Thin arrows at each step showing direction & magnitude of movement (gradients).

3D Straightening Trajectories with Gradient Arrows

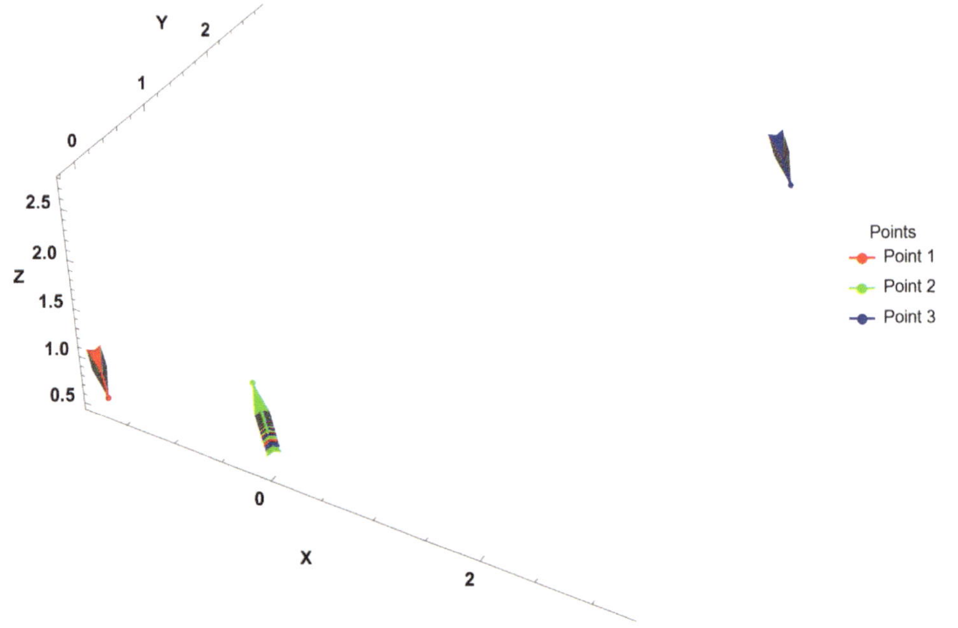

Fig. 3: Projects points onto XY plane.
Draws their movement as colored lines.
Overlays black contour lines of the deviation function's value around the points — showing how "far" configurations are from perfect straightening in 2D.

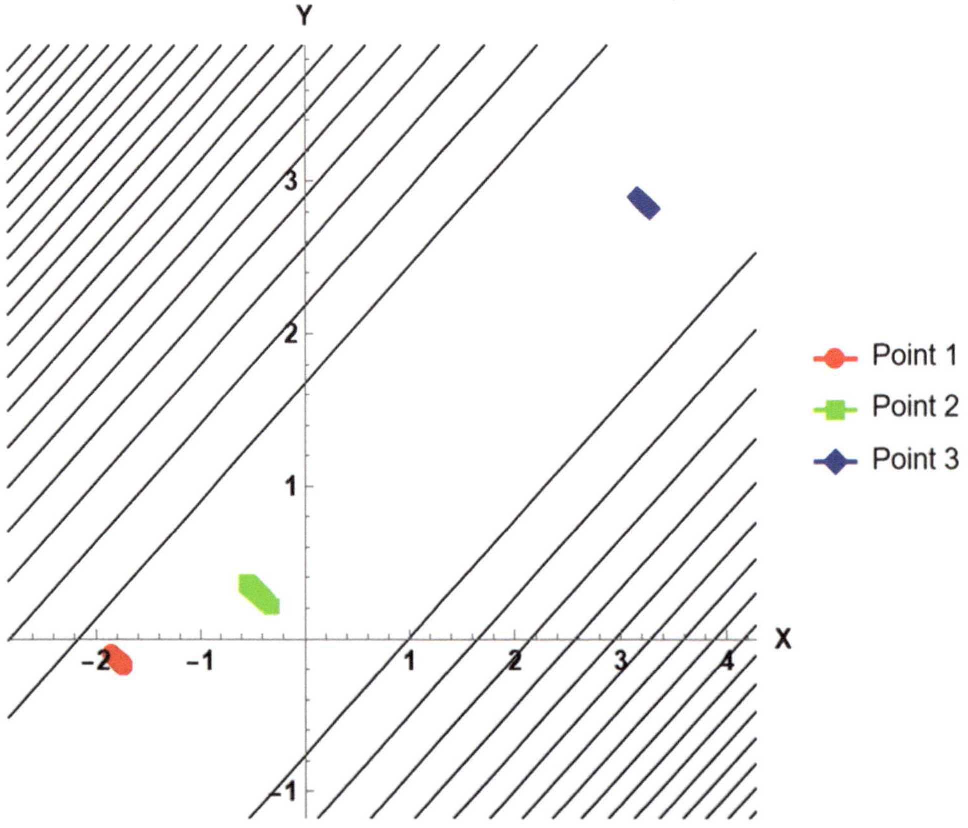

IX. Generalization

let's generalize the straightening algorithm to **multiple 3D points** under force-based correction, keeping the physics-based and tolerance-aware constraints.

Problem Setup (Generalized)
Given:
A set of N points $\{P_i\}_{i=1}^{N} \subset \mathbb{R}^3$
Each has:

- Target position: $P_i^T = (x_i, y_i, z_i)$
- Current position: $P_i^0 = \left(x_i^{(0)}, y_i^{(0)}, z_i^{(0)}\right)$
- Tolerance per axis: $\tau_i = \left(\tau_{ix}, \tau_{iy}, \tau_{iz}\right)$

Material and impact parameters:

- Yield Stress YS
- Ultimate Tensile Strength UTS
- Elongation A_{mat}
- Impact Strength IS
- Impact area S
- Velocity v

Iterative Straightening Algorithm — Generalized

1. Initialize

For each point i∈{1,…,N}

Define:$\epsilon_i = \sqrt{\tau_{ix}^2 + \tau_{iy}^2 + \tau_{iz}^2}$
Set: $P_i^0 \in \mathbb{R}^3$ (initial position)
Set damping factor λ∈(0,1) initial step size α_0 , and iteration count k=0.

In the context of the iterative straightening algorithm, the damping factor λ∈(0,1) is a scalar multiplier used to gradually reduce the step size (i.e., the magnitude of corrections or applied forces) in each iteration.

It helps the algorithm:

- Stabilize: Avoid overshooting or oscillations as the nodes get close to their targets
- Converge smoothly: Makes the final steps smaller and more accurate
- Mimic physical dissipation: Real systems lose energy over time (like friction or internal material damping)

2. Iteration Loop

For iteration k, repeat until all points are within tolerance:

(a) Compute deviation vectors: $\Delta P_i^{(k)} = P_i^{(k)} - P_i^T \in \mathbb{R}^3$

(b) Deviation norm: $\delta_i^{(k)} = \left\|\Delta P_i^{(k)}\right\|_2 = \sqrt{\left(x_i^{(k)} - x_i\right)^2 + \left(y_i^{(k)} - y_i\right)^2 + \left(z_i^{(k)} - z_i\right)^2}$

(c) Convergence check:

For each i: $If\ \delta_i^k \leq \epsilon_i, mark\ point\ as\ converged$

3. Force Estimate and Correction

Only for unconverged points i, apply:

(a) Estimated required impact force:

$$F_i^{(k)} = \frac{IS \cdot S}{\delta_i^{(k)} + \epsilon}$$

ε is a small constant to avoid division by zero

(b) Scaled correction vector "C":

Let: $\alpha_k = \alpha_0 \cdot \lambda^k$

Then: $c_i^{(k)} = -\alpha_k \cdot \frac{\triangle P_i^{(k)}}{\delta_i^{(k)} + \epsilon}$

(c) Update point positions:

$$P_i^{(k+1)} = P_i^{(k)} + C_i^{(k)}$$

4. Convergence

Terminate when:

$$\forall i \in \{1, \dots, N\}, \delta_i^{(k)} \leq \epsilon_i$$

Matrix Formulation (Optional Compact View)

Let:

$P^{(k)} \in \mathbb{R}^{N \times 3}$: matrix of current positions

$P^T \in \mathbb{R}^{N \times 3}$: matrix of current positions

$$\triangle^{(k)} = P^{(k)} - P^T \in \mathbb{R}^{N \times 3}$$

Define norm vector: $\delta_i^{(k)} = \left\| \triangle_i^{(k)} \right\|_2, for\ i = 1, \dots, N$

Correction matrix: $C_i^{(k)} = \begin{cases} -\alpha_k \cdot \frac{\triangle_i^{(k)}}{\delta_i^{(k)} + \epsilon}, if \delta_i^{(k)} > \epsilon_i \\ 0,\ otherwise \end{cases}$

Update: $P^{(k+1)} = P^{(k)} + C^{(k)}$

<u>Repeat until all</u> $\delta_i^{(k)} \leq \epsilon_i$

Optional Extensions

- Weighting impacts by material anisotropy or point stiffness
- Time-dependent control: allow α to vary per point based on curvature
- Multi-impulse control: Apply to selected subsets based on ranking δ_i
- Control point re-placement: dynamically update external control points like actuators or impactors

Straightening Order Strategy

We want to prioritize which point(s) to correct at each iteration. This means choosing a subset of the N points to receive force or correction based on some metric.

Step 1: Define Priority Metric

The most natural metric is the deviation magnitude:

$$\delta_i^{(k)} = \left\| P_i^{(k)} - P^T \right\|$$

This tells how far point i is from its desired (tolerance-based) location.

Step 2: Sort the Points by Deviation

At iteration k, define the priority list:

$$\pi^{(k)} = argsort\left(\left\{\delta_i^{(k)}\right\}_{i=1}^{N}\right)$$

This produces an ordered list $\pi^{(k)} = [i_1, i_2, \dots, i_N]\ where\ \delta_{i1}^{(k)} \geq \delta_{i2}^{(k)} \geq \cdots \geq \delta_{iN}^{(k)}$

So point i_1 is the most deviated, and point i_N is closest to tolerance.

Step 3: Straightening Selection Rule

Now, choose how many top-ranked points to straighten at each iteration.

Let's define:

- M_k : number of points to straighten at iteration k
- You can set:
- Fixed: M_k=M (e.g. always straighten top 3)
- Dynamic: $M_k = min\left\{i: \delta_i^{(k)} > \epsilon_i\right\}$
- Then: $S^{(k)} = \left\{\pi_i^{(k)}, \dots, \pi_{M_k}^{(k)}\right\}$

is the set of points you will apply correction to at iteration k.

Final Update Rule with Straightening Order

For each i∈{1,…,N}

$$C_i^{(k)} = \begin{cases} -\alpha_k \cdot \dfrac{P_i^{(k)} - P_i^T}{\left\| P_i^{(k)} - P_i^T \right\| + \epsilon}, if\ i \in S^{(k)} \\ 0,\ otherwise \end{cases}$$

$$P^{(k+1)} = P_i^{(k)} + C_i^{(k)}$$

Then increase k→k+1 and repeat.

Strategy Options

Strategy	Type Rule
Greedy	Straighten only the most deviated point: Mk=1
Batch	Always straighten top M points
Adaptive	Straighten all points with $\delta_i^{(k)} > \epsilon_i$
Weighted	Apply larger α_k to worse points (soft prioritization)
Force Budget	Distribute limited total force among top deviated nodes

Why Ordering Matters

- Reduces unnecessary corrections where convergence is already near
- Prevents interfering impulses from nearby control points
- Makes force application more targeted, efficient, and physically plausible

Note:

The Difference: k vs. (k)

✅ k — Index / Iteration Counter

Plain k is used when we refer to:

- Which iteration we're in
- The current loop step
- General iteration count

✅ (k) — Superscript Notation (State at Iteration k)

This is the value of a variable after the k-th iteration. It's common in numerical methods and optimization to use superscripted parentheses for time-step or iteration indices. It does NOT mean exponentiation.

Example (see also fig. 1)

controlPoints = {{1, 1, 1}, {2.5, 2.5, 1}, {3.5, 1, 1.5}};

epsilon = 0.1;

These are 3D points in space (with z-coordinates), acting as "sources" of effort.

The field around each point increases sharply near it and falls off farther away.

ε is a smoothing parameter that ensures the function is well-behaved and avoids singularities when a point gets too close to a source.

Fig. 1: The following visual illustrates the spatial effort field h(x) and its gradient over a 2D slice of the 3D domain, helping to intuitively understand the directional forces used during iterative straightening.

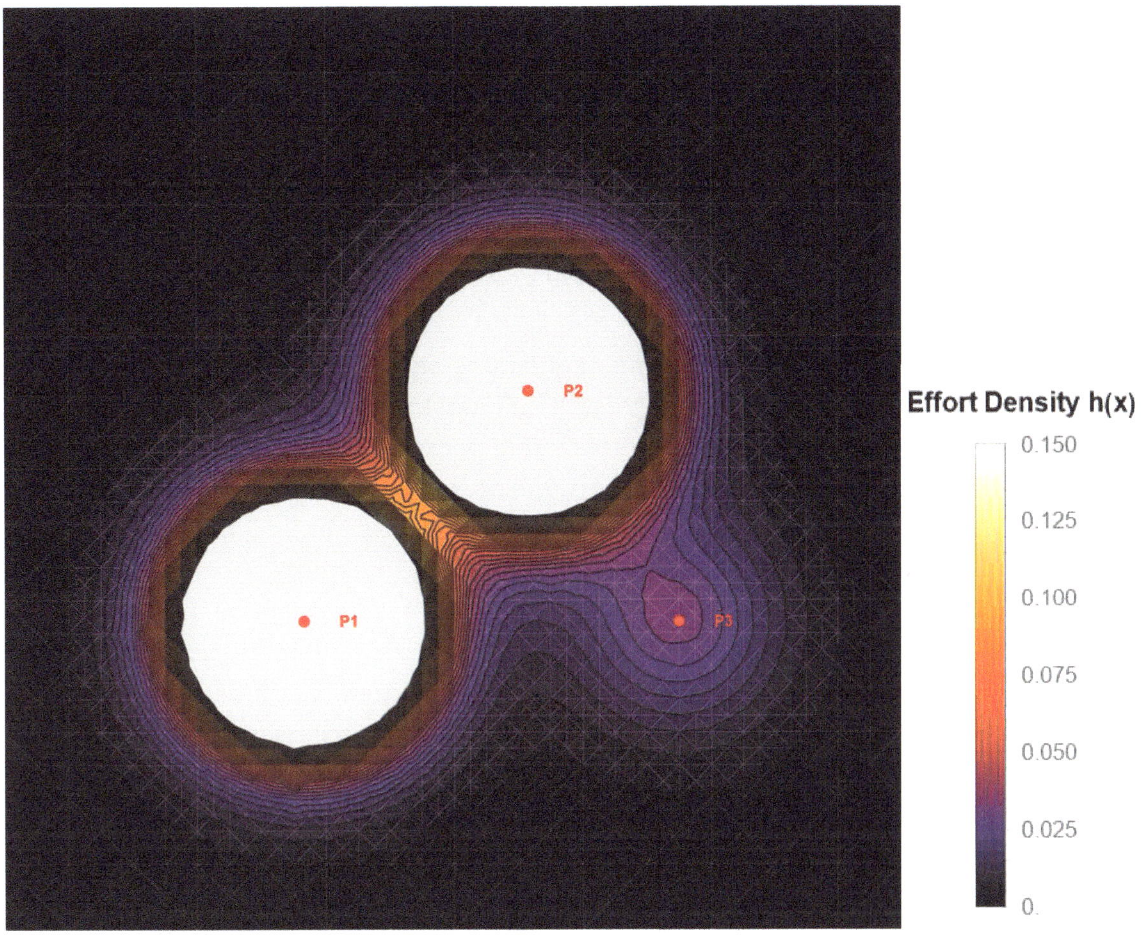

Physical Meaning in the Straightening Algorithm

Region in Plot	What It Tells You
High effort (bright)	Material needs strong correction — apply or maintain force
Low effort (dark)	Material near equilibrium — can be marked as converged
Multiple bright spots	Multiple control points influencing different parts of the body
Flat valleys between peaks	Transition zones — moderate deformation, partial convergence

Component	Interpretation
Control Points Pi	Fixed "anchors" generating effort (or resistance). They guide the deformation process.
Effort Field h(x)	A scalar field quantifying how difficult it is to be in a certain region — like a "resistance map". The goal is to minimize exposure to high-effort zones.
Gradient of h(x)	This gives the direction of steepest increase in effort. To "straighten" a path or configuration, you would move opposite this direction — downhill.
Contour Surfaces (isosurfaces of constant h(x))	These are boundaries of equal effort. A path or surface should ideally slide down through these contours like a marble rolling down hills.
Smoothing Parameter ε	Controls how sharp or smooth the influence of each point is — sharper values mean a more "localized" resistance.

IX.1 Fractal-Controlled Straightening

Objective:

Generalize the straightening algorithm further by introducing **fractal-based control point structures** (see fig. 1). These simulate complex or recursive patterns of influence, ideal for modeling multi-scale deformation (e.g., leaf veins, bones, or material lattices).

Let each point Pi(k) be corrected in the direction of **lowest effort**, i.e., opposite to ∇h(x), like a marble sliding downhill:

Fractal Straightening Iteration (conceptual steps)

1. Initialize all point positions $P_{i(0)}$
2. For each iteration k:
    - For each point:

- Evaluate ∇h(x) at current position
- Move opposite to gradient:

$$P_i^{k+1} = P_i^k - \alpha_k \cdot \nabla h(P_i^k)$$

- Reduce step size:

$$\alpha_k = \alpha_0 \cdot \lambda^k$$

- **Convergence**: stop if all points within tolerance ε_i

Fig. 1: Sierpiński triangle fractal algorithm 2D

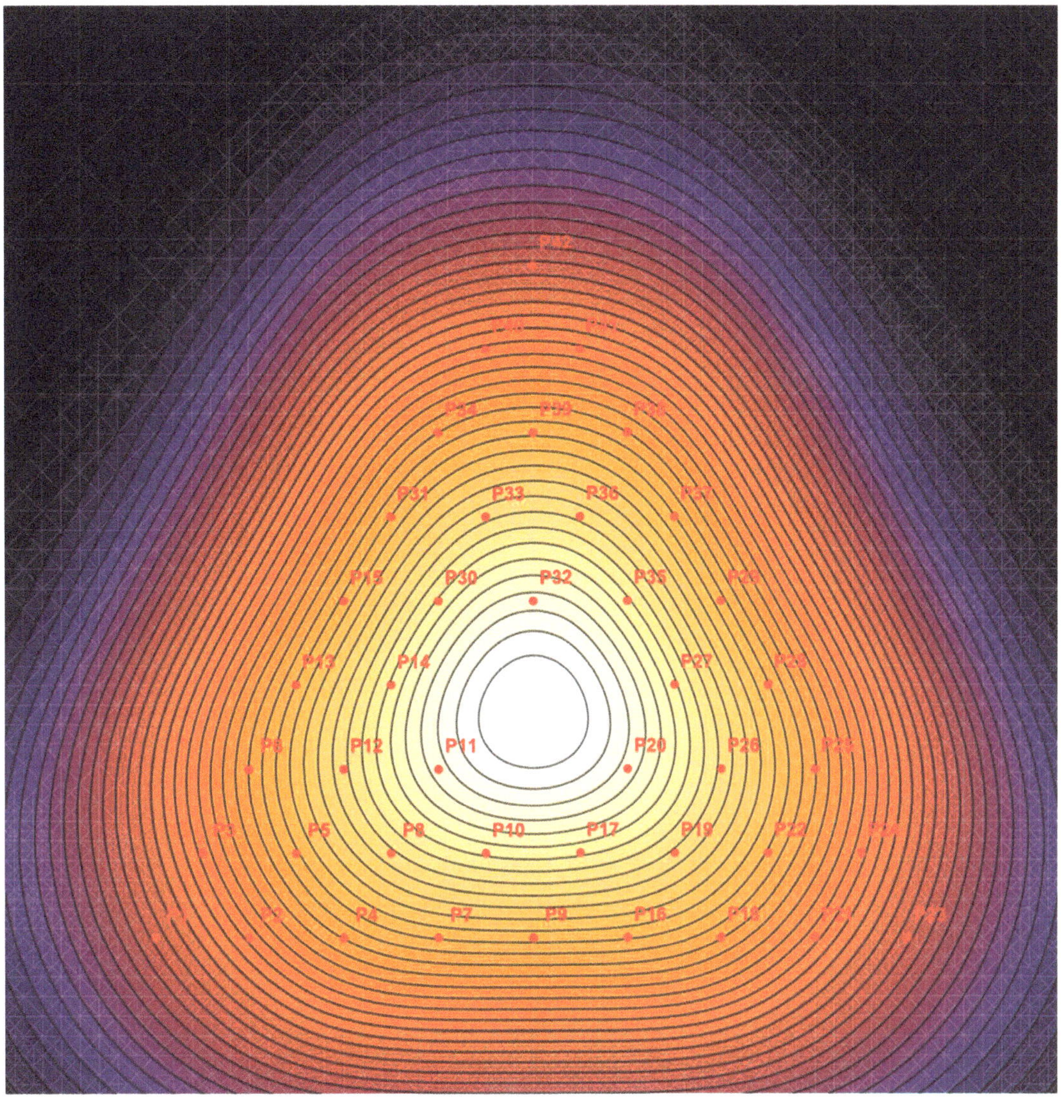

What the Graphic Shows

The graphic visualizes an "effort field" generated by multiple control points arranged in a fractal pattern on a 2D plane (z=0), which is interpreted as a 3D scenario with zero height.

- Effort Field (h(x, y))

  This scalar field quantifies the *"resistance"* or *"difficulty"* at any point (x,y) in the plane, influenced by all the control points. Brighter colors correspond to regions of high effort (strong resistance), and darker colors show areas near equilibrium.

- Contour Plot (Background Colors)

  The contour plot uses smooth color gradations to map the values of the effort field h(x,y)

Peaks (bright spots) correspond to the control points exerting the strongest local influence.

Valleys and dark areas correspond to low effort zones, where the field is near equilibrium.

- Red Points with Labels

  Each red dot marks a control point—a source of "effort"—and is labeled as P_i. These points act like "anchors" or "force application sites" that influence the entire field.

How Control Points Are Generated Procedurally

Instead of defining control points manually (like {1,1,1},{2.5,2.5,1},{3.5,1,1.5}, the control points are generated using the **Sierpiński triangle fractal algorithm**:

What is the Sierpiński Triangle?

- A classic fractal formed by recursively subdividing a triangle into smaller triangles and removing the central one.
- It exhibits self-similarity—each smaller part looks like the whole, repeated infinitely.
- The fractal is a set of points arranged in a highly regular, nested pattern, with an intricate geometric structure.

Inputs: Three vertices p_1,p_2,p_3 of a triangle and a recursion depth.

Process:

- At each recursion, it calculates midpoints of each side: m_{12}, m_{23} ,m_{31}.
- It splits the triangle into 3 smaller triangles (excluding the central one).
- Recursively calls itself on each smaller triangle until depth reaches 0.

Output: A list of points corresponding to the vertices of the smallest triangles at the final recursion level.

Why use fractal control points?

- They generate a rich, hierarchical distribution of control points, simulating complex natural structures or stress distributions.
- The fractal pattern provides multi-scale influence — large-scale structure with small-scale details.
- This leads to an effort field with multiple bright peaks, valleys, and transition zones — a more complex and realistic scenario than just a few isolated points.

What the Graphic Tells Physically

- High-effort bright spots: Areas near fractal control points needing strong correction forces.
- Multiple bright spots: Reflect influence from many fractal points, modeling complex behavior.
- Gradient vectors: Indicate where the "effort" pulls strongest; moving opposite reduces effort.
- Flat valleys: Regions where effort is low — closer to desired equilibrium or tolerance.

In summary:

- The fractal algorithm creates the control points.
- These points are sources of effort that shape the overall field.

- The visualization maps how effort varies across the 2D plane, guiding where corrections would be needed in a physical system.
- The gradient vectors show the direction corrections should be applied to minimize effort.

IX.2 Lyapunov Stability

Let's relate this to the iterative straightening algorithm or effort-gradient-based system.

The system evolves as:

$$P^{k+1} = P^k - \alpha_k \cdot \frac{\Delta P^k}{\|\Delta P^k\| + \epsilon}$$

Let's treat P^k as a discrete-time dynamical system.

Can it be Lyapunov-stable? (see also fig. 2 and 3)

Yes — if we can construct a Lyapunov function like

$$V(P^k) = \sum_i \|\Delta P_i^k\|^2$$

Then the update rule ensures:

$$V(P^{k+1}) < V(P^k) \; as \; long \; as \; \alpha_k > 0$$

Because:

- You're descending the gradient of the error.
- The correction vector reduces the deviation.
- Over iterations, the total deviation $\|\Delta P^k\| \to 0$

✅ This implies asymptotic stability of your iterative correction.

✅ The equilibrium point is when all $P_i = P_i^T$, i.e., zero error.

Why It Matters

- Lyapunov analysis gives guarantees about your algorithm.
- It ensures that the system won't oscillate or diverge.
- In physical terms: it mimics energy dissipation, as in real deformation systems.

The Lyapunov stability explanation refers specifically to the "Iterative Straightening Algorithm — Generalized" described in Chapter IX, where points in 3D are corrected iteratively using a force-based model.

Here's the key part of the algorithm: defining a recursive update of the form:

$$P^{k+1} = P^k - \alpha_k \cdot \frac{\Delta P^k}{\|\Delta P^k\| + \epsilon}$$

This is a **discrete-time system**, where each point moves closer to its target position over iterations.

Why It Applies:

This structure is formally equivalent to a discrete-time gradient descent method. In such systems, Lyapunov stability is analyzed using an energy-like function. In the present case, we can define:

$$V^{(k)} = \sum_{i=1}^{N} \left\| P_i^{(k)} - P_i^T \right\|^2$$

Term-by-Term Explanation

Symbol	Meaning
N	Number of points being corrected
$P_i^{(k)}$	The position of point iii at iteration k
P_i^T	The target (desired) position of point i
$\left\| P_i^{(k)} - P_i^T \right\|$	The Euclidean distance (deviation) between the current and target position of point i
$\|\cdot\|^2$	Squared norm (i.e., square of the distance

What Does $V^{(k)}$ Represent?

It's a global error metric (or energy-like function):

- Measures the total squared deviation of all points from their targets.
- Always non-negative (since it's a sum of squares).
- Equals zero only when all points have reached their targets exactly (i.e., perfect convergence).
- Decreases with each iteration if the algorithm is working correctly.

- Measures total deviation (energy or "effort") across all points.
- Decreases with each iteration, as points move toward their target.
- Reaches minimum (zero) when all points are within their respective tolerances.

This matches the behavior of a Lyapunov function — confirming asymptotic stability of your straightening process.

Why Is It a Lyapunov Function?

In control theory, a Lyapunov function is used to prove stability of a system.

If $V^{(k+1)} < V^{(k)}$ for all k, then the system is asymptotically stable — meaning the algorithm converges to the desired state (in our case, the straightened positions). Since we're applying a damping factor λ<1, the step size decreases, ensuring that corrections get smaller and $V^{(k)} \to 0\ ask \to \infty$.

This **does NOT apply to** the fractal visualization algorithm (the one based on the Sierpiński triangle), because:

- That code does not perform iteration or correction.
- It is a static visualization of a scalar effort field and its gradient.
- There is no dynamic system or state evolution to analyze.

Fig. 2: Plots how **V(k)** decreases → showing stability and convergence.

- targetPoints = {{1, 1, 0}, {2, 2, 0}, {3, 1, 0}};
- initialPoints = {{0, 2, 0}, {2.5, 2.5, 0}, {4, 2, 0}};

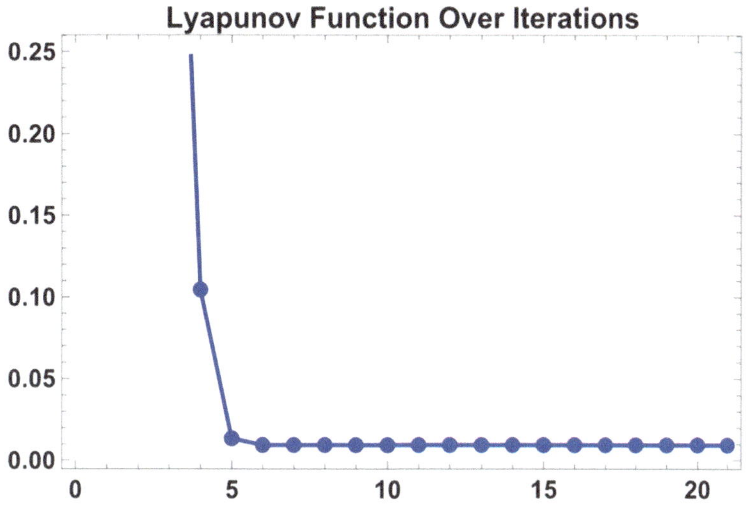

Fig.: 3D Trajectories of Control Points in Straightening Algorithm
targetPoints = {{1, 1, 0}, {2, 2, 0}, {3, 1, 0}};
initialPoints = {{0, 2, 0}, {2.5, 2.5, 0}, {4, 2, 0}}

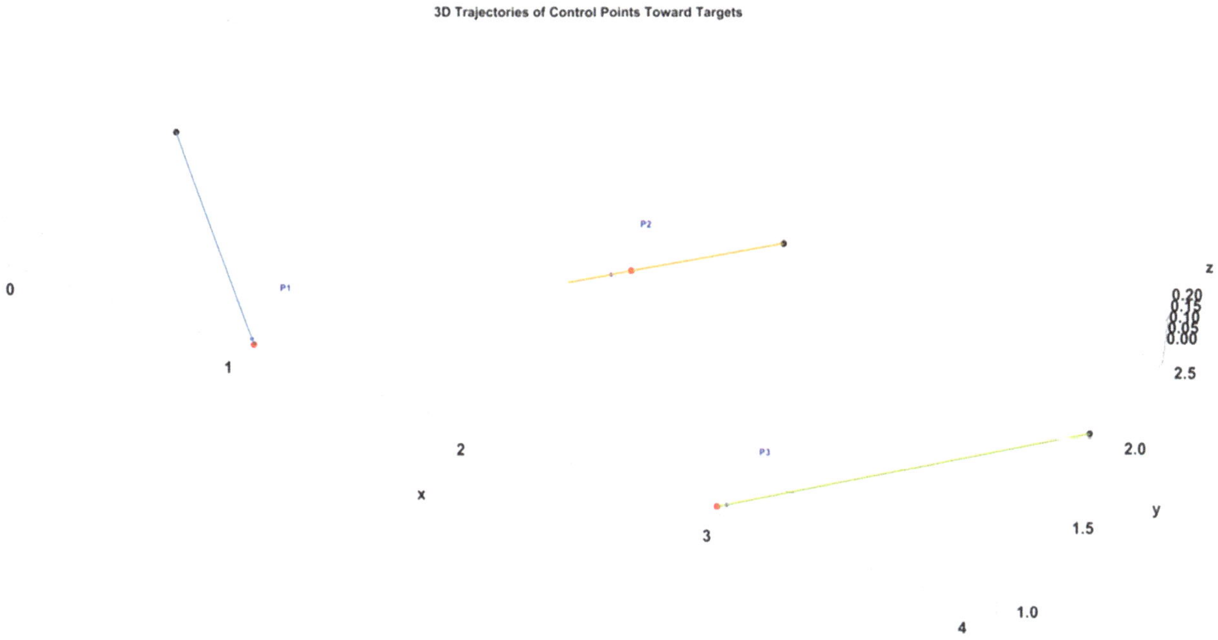

Colored 3D lines: Each shows the path of a point moving toward its target.
Black dots: Initial positions. Red dots: Final positions after convergence.
Blue dots + labels: Target positions (with P1, P2, P3 labels).

X. Fictitious Example – Straightening Process

Parameters and Conditions

1.Initial Coordinates $P_i^{(0)}$

The starting 3D positions of the points represent the current bent or misaligned shape:

Point i	$x_i^{(0)}$	$y_i^{(0)}$	$z_i^{(0)}$
1	1.0	1.0	1.0
2	2.5	2.5	1.0
3	3.5	1.0	1.5
4	2.0	0.5	2.0

2. Target Coordinates P_i^T

The desired straightened positions, often aligned along a reference axis or smooth curve:

Point i	x_i^T	y_i^T	z_i^T
1	1.0	1.0	1.0
2	2.5	2.0	1.2
3	3.5	1.0	1.5
4	2.0	1.0	1.8

3.Tolerances $\tau_i = (\tau_{ix}, \tau_{iy}, \tau_{iz})$

Allowed deviation per axis for each point; defines convergence thresholds:

Point i	τ_{ix}	τ_{iy}	τ_{iz}	Overall Tolerance$\epsilon_i = \sqrt{\tau_{ix}^2 + \tau_{iy}^2 + \tau_{iz}^2}$
1	0.05	0.05	0.05	0.0866
2	0.05	0.05	0.05	0.0866
3	0.05	0.05	0.05	0.0866
4	0.05	0.05	0.05	0.0866

Interpretation: When the Euclidean distance between the current position and target is less than ϵ_i, point i is considered straightened.

4. Material and Impact Parameters

These characterize the physical properties affecting the straightening forces and corrections:

Parameter	Symbol	Value	Physical Meaning
Yield Stress	YS	250MPa	Stress at which material begins to deform plastically
Ultimate Tensile Strength	UTS	450MPa	Maximum stress material can sustain before failure
Elongation	A_{mat}	20%(0.2)	Material's ductility or ability to elongate without failure
Impact Strength	IS	$100J/m^2$	Energy absorbed per unit area during impact
Impact Area	S	$0.005m^2$	Area over which force is applied
Velocity	v	2m/s	Speed of the impactor or correction device

5. Force Estimation for Corrections

At iteration, the deviation norm for point i is:

$$\delta_i^{(k)} = \left\| P_i^{(k)} - P^T \right\|$$

The corrective impact force applied is modeled as:

$$F_i^{(k)} = \frac{IS \cdot S}{\delta_i^{(k)} + \epsilon}$$

6. Impulse and Correction Vector

- The correction step at iteration k moves point i by:

$$c_i^{(k)} = -\alpha_k \cdot \frac{\Delta P_i^{(k)}}{\delta_i^{(k)} + \epsilon}$$

where:

- $\Delta P_i^{(k)} = P_i^{(k)} - P_i^T$
- $\alpha_k = \alpha_0 \times \lambda^k$
- $\lambda \in (0,1)$ is the damping factor
- α_0 is is the initial step size.

7. Iterative Update Rule

The new position is:

$$P_i^{(k+1)} = P_i^{(k)} + C_i^{(k)}$$

This iterative correction continues until: $\forall i \in \delta_i^{(k)} \leq \epsilon_i$

Summary

- Initially, points may be far from their targets, exceeding tolerance bounds.
- The algorithm applies physically-informed corrective forces, scaling with impact strength and deviation magnitude.
- Over iterations, corrections become smaller due to damping and proximity to target.
- When points fall within the tolerance sphere (radius ϵ_i), they are considered fully straightened.
- The material parameters (yield stress, elongation, impact strength) guide realistic correction magnitudes respecting physical limits.
- This ensures the straightening is both effective and safe, avoiding damage or over-correction.

The order of points during impulse application in the straightening algorithm is crucial for efficient, stable, and physically realistic correction.

How the Straightening Order is Determined

1. Priority Metric: Deviation Magnitude

At each iteration k, compute for every point i:

$$\delta_i^{(k)} = \left\| P_i^{(k)} - P^T \right\|$$

This deviation is how far the current point is from its target position.

2. Sorting Points by Deviation

The points are ranked by their current deviation in descending order:

$$\pi^{(k)} = argsort\left(\delta_1^{(k)}, \delta_2^{(k)}, \dots, \delta_N^{(k)}\right)$$

$$where\ \delta_{\pi_1}^{(k)} \geq \delta_{\pi_2}^{(k)} \geq \cdots \geq \delta_{\pi_N}^{(k)}$$

The point with the largest deviation π_1 is considered the highest priority for correction.

The point closest to its target is the lowest priority.

3. Selection of Points to Correct per Iteration

- correct:
    - Only the top 1 point (Greedy approach): Correct the most deviated point only.
    - Top M points (Batch approach): Correct the top M deviated points simultaneously.
    - All points above tolerance (Adaptive approach): Correct every point with deviation exceeding its tolerance ϵ_i.

4. Applying Corrections in Order

- The impulses are applied starting with the most deviated point.
- This approach reduces unnecessary corrections to already nearly-straight points.
- Prevents multiple conflicting corrections that could cause oscillations or instability.
- Mirrors a physical scenario where you would impact the "worst bent" spot first, then proceed to the next.

Why this ordering matters

- Efficiency: Focus computational effort where it's most needed.
- Stability: Avoid multiple competing impulses that could overshoot or cancel.
- Physical realism: Impacts happen sequentially on the most critical bends.
- Convergence: Ensures smoother, faster straightening.

Summary Table

Step in Iteration k	Description
1	Calculate deviations $\delta_i^{(k)}$
2	Sort points descending by $\delta_i^{(k)}$
3	Select subset $S^{(k)}$ to correct (top M or all over tolerance)
4	Apply impulses $C_i^{(k)}$ in order of descending deviation
5	Update positions $$P_i^{(k+1)} = P_i^{(k)} + C_i^{(k)}$$

What is M?

- M is the number of points you decide to straighten (correct) simultaneously at each iteration.

How to choose M:

- **M=1 (Greedy strategy):**

  Only the **single most deviated point** is corrected in each iteration. This is the most cautious approach, focusing effort on one point at a time.

- **M>1 (Batch strategy):**

Correct the **top M points** with the largest deviations simultaneously. For example, M=3 means the three points with the largest deviations get corrected in each iteration.

- **M=N (Adaptive or Full batch):**

All points with deviation exceeding their tolerance are corrected simultaneously.

Example

Tolerances: for all 4 points in x,y,z directions: 0.05,0.05,0.05

Initial points

Point	x	y	z
1	1	1	1
2	2.5	2.5	1
3	3.5	1	1.5
4	2	0.5	1

Target points

Point	x	y	z
1	1	1	1
2	2.5	2	1.2
3	3.5	1	1.5
4	2	1	1.8

Final points after 2 iterrations

Point	x	y	z
1	1	1	1
2	2.5	2.035	1.185
3	3.5	1	1.5
4	2	0.964	1.814

The following figures (fig. 1 and 2) graphically depict the above iterrations.

Fig. 1:

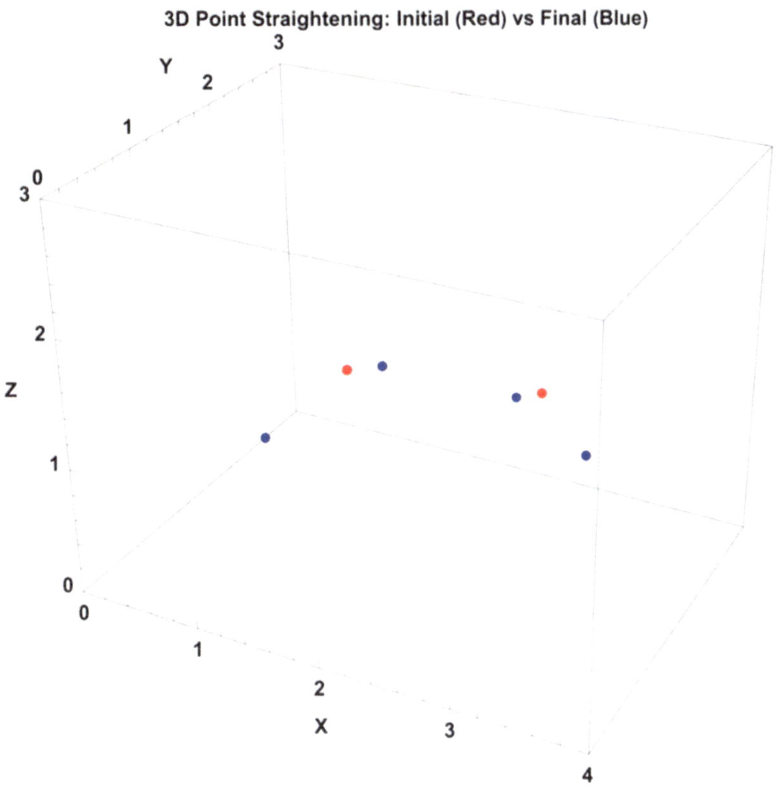
3D Point Straightening: Initial (Red) vs Final (Blue)
Y
Z
X
0
1
2
3
4

Fig. 2:

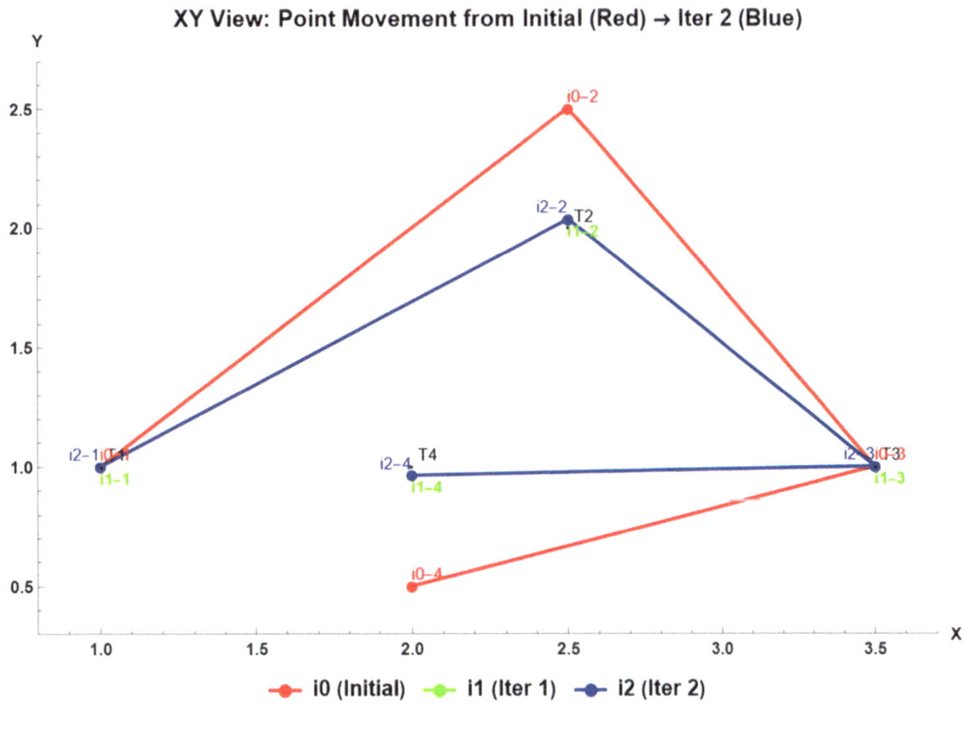
XY View: Point Movement from Initial (Red) → Iter 2 (Blue)
Y
X
2.5
2.0
1.5
1.0
0.5
3.0
3.5
i0–2
i2–2
T2
i2–1
i1–1
i2–4
T4
i1–4
i1–3
i0–4
i0 (Initial)
i1 (Iter 1)
i2 (Iter 2)

Strategy Summary (see also table 1):

- ✅ Batch approach: ✔ Used here (M = 2)
- ❌ Greedy approach: Would be M = 1

✅ What is happening in your Batch approach (M = 2)?

- It selects only the top 2 deviated points, even if more than 2 points are still not converged.
- If only 2 points are outside their tolerances, then Batch(M=2) = Adaptive.
- If more than 2 are outside tolerance, Batch(M=2) $\subset$ Adaptive (subset of what adaptive would correct).

Deviations:

- Point 1: Perfect match → deviation = 0 → ✅ converged
- Point 2: deviation = $\sqrt{[(0)^2 + (0.5)^2 + (0.2)2]} = \sqrt{(0.29)} \approx 0.5385$
- Point 3: Perfect match → deviation = 0 → ✅ converged
- Point 4: deviation = $\sqrt{[(0)^2 + (0.5)2 + (0.2)^2]} = \sqrt{(0.29)} \approx 0.5385$

So, at iteration 0:

- Only Points 2 and 4 exceed tolerance → Adaptive set = {2, 4}
- Batch with M = 2 will also choose top 2 deviated points → also {2, 4}

✅ Conclusion: In this specific scenario, where only two points are outside their tolerance, the Batch approach with M = 2 behaves identically to the Adaptive approach.

However:

- This equivalence is <u>coincidental</u> due to the problem data.
- If, say, 3 points had deviations above their ϵ_i, then:
  - Batch(M=2) would correct only the top 2,
  - Adaptive would correct all 3

Table 1: a possible comparison of the different approaches

Point	Greedy Init	Greedy Final	Greedy Improvement	Batch Init	Batch Final	Batch Improvement	Adaptive Init	Adaptive Final	Adaptive Improvement
1	0.0000	0.0000	0.0000	0.0000	0.0000	0.0000	0.0000	0.0000	0.0000
2	0.5385	0.2477	0.2908	0.5385	0.0385	0.5000	0.5385	0.0385	0.5000
3	0.0000	0.0000	0.0000	0.0000	0.0000	0.0000	0.0000	0.0000	0.0000
4	0.5385	0.0385	0.5000	0.5385	0.0385	0.5000	0.5385	0.0385	0.5000

(Comparison) Table 2: Without vs With Physical Constraints

Point	Initial Position	Target Position	Final (Simple)	Final (Physical)	Status (Physical)
1	1.000 1.000 1.000	1.000 1.000 1.000	1.000 1.000 1.000	1.000 1.000 1.000	Converged
2	2.500 2.500 1.000	2.500 2.000 1.200	2.500 2.036 1.186	2.500 2.500 1.000	Fractured
3	3.500 1.000 1.500	3.500 1.000 1.500	3.500 1.000 1.500	3.500 1.000 1.500	Converged
4	2.000 0.500 2.000	2.000 1.000 1.800	2.000 0.964 1.814	2.000 0.500 2.000	Fractured

Fracture at iteration 0 — that means right at the start, the points (of the imaginary body) violate the physical constraints; the very first iteration already triggers fracture conditions. This means the initial state of the points, combined with the physical parameters, implies the material is already beyond safe limits before any correction is applied.

In other words, the *initial configuration is physically "fractured" or unstable* under the current constraints (see table 2).

Commentary on the Meaning of "Fracture at Iteration 0" – critical clarification in any scientific or engineering discussion involving simulation and abstraction.

The phrase:

"Fracture at iteration 0" does not imply that a real material object has fractured. Rather, it refers to a hypothetical violation of physical constraints in a mathematical model.

Physical Infeasibility vs. Algorithmic Utility

- In the present algorithm, *fracture* simply means:

The initial conditions, when evaluated against yield strength, impact tolerance, or stress limits, imply that the modeled material would not survive the deformation if it were real.

- This does not mean a real object is being tested.
- The algorithm does not simulate actual materials directly — it provides a framework for reasoning about deformation, correction, and convergence using abstract control points.

Why This Algorithm Is Valuable

Despite not representing a physical object:

1. It provides insight into:
    - How deviations evolve under correction.
    - The role of damping and force scaling.
    - Convergence behavior under constraints.
2. It is adaptable:
    - Physical parameters (impact strength, yield stress, etc.) can be tuned to explore various material behavior scenarios.
3. It informs decision-making:
    - Even if physical violation occurs (e.g., "fracture"), the model helps identify instability or design limits.
4. It generalizes well:
    - The same principles apply to many problems in structural optimization, robotics, soft body mechanics, and even computer graphics — not only physical straightening.

Final Interpretation: Although this algorithm models a fictitious body that would not physically survive the initial conditions, it remains a mathematically sound, physically inspired, and computationally useful tool for analyzing convergence, stability, and correction dynamics in spatial configurations. It's a testbed, not a real object.

XI. Conlusions and Perspectives

As we reach the final chapter of this work, it is essential to move beyond technical formulations and reflect on the broader significance of the tools, ideas, and future directions emerging at the intersection of **artificial intelligence (AI)**, **mathematics**, and **engineering knowledge**, particularly in fields such as **metallurgy** and **structural mechanics**.

AI in Computational Mechanics and Metallurgy

Artificial Intelligence has undoubtedly begun to reshape the landscape of engineering analysis. Once confined to rule-based systems or empirical data fitting, AI now offers advanced capabilities such as real-time simulation, pattern recognition in complex material behavior, and automated control of mechanical processes. For instance, in the **straightening of metallic bodies**—a process long reliant on empirical adjustments and expert intuition—AI systems can now learn from historical correction paths, mechanical feedback, and material memory to optimize tool paths and force applications. Modern AI, especially when combined with physics-informed models, is capable of predicting springback, plastic deformation limits, and even microstructural effects after deformation.

However, these systems are not self-aware or omniscient. They **do not "understand" metallurgy**—they model it. Behind every intelligent system lies a lattice of mathematical formulations and approximations constructed by human insight. Without deep knowledge of yield criteria, strain hardening behavior, residual stress evolution, and thermal-mechanical coupling, AI is simply manipulating data. In straightening applications, where decisions depend on subtle material variations and decades of hands-on experience, **AI cannot replace metallurgists**—but it can support and amplify their decision-making power.

Mathematics: The Universal Interface

Mathematics remains the one language both humans and machines must speak fluently to collaborate effectively. Whether encoding a stress-strain relationship, defining an optimization functional, or expressing convergence criteria, **mathematics is not a barrier but a bridge**. It allows the abstraction of physical phenomena into forms that algorithms can manipulate, simulate, and learn from.

In structural mechanics and metallurgy, governing equations—elasticity, plasticity, thermomechanical coupling—are all mathematical in nature. Whether using finite elements, neural networks, or hybrid solvers, these tools are **mathematical translators** of the physical world. As such, **advancing mathematical understanding** remains key—not only for developing better algorithms but also for **understanding where algorithms fail**.

Direct Programming and Transparency

In a time when software grows increasingly opaque, the value of **direct programming**—where one implements algorithms at the most granular level—cannot be overstated. While AI models may automate routine tasks, they often obscure the reasoning behind decisions. Direct code, by contrast, forces clarity of logic, exposes numerical instability, and encourages a deeper relationship between algorithm and phenomenon.

Engineers and scientists who write their own solvers, even at the cost of computational efficiency, develop a **critical understanding** of the assumptions, simplifications, and limitations behind every result. Particularly in safety-critical applications like structural straightening or residual stress relief in load-bearing components, blind trust in commercial solvers or black-box AI is a liability. Direct programming restores control and accountability—values that AI must learn to respect.

The Irreplaceable Role of Human Expertise

Throughout this book, we have explored methods grounded in finite element analysis, nonlinear solving strategies, and numerical stability. But perhaps the most important message is this: **knowledge is not stored in code—it is embodied in experience**.

In metallurgy, where materials behave differently across batches, temperatures, and geometries, **no algorithm can replace the nuanced judgment of a seasoned engineer**. AI can suggest corrections, highlight anomalies, or propose novel configurations—but it is **human knowledge** that validates, interprets, and ultimately applies them in real-world contexts.

Just as a skilled metallurgist understands that every beam bent beyond its yield has a story in its microstructure, the future of simulation and automation must respect that **true intelligence is interpretative, not just predictive**.

Looking Forward

The path forward is neither purely manual nor purely automated—it is hybrid. The convergence of AI, advanced mathematics, and programmable tools offers unprecedented capability, but only when wielded with **human insight, domain expertise, and ethical responsibility**.

- Let AI handle the repetition.
- Let mathematics encode the truth.
- Let programming give us precision.
- But let **humans guide the purpose**.

This book closes with a challenge, not a conclusion: to build tools that reflect not only our knowledge, but our wisdom. In metallurgy, in mechanics, and beyond, that will make all the difference.

"This result is too beautiful to be false; it is more important to have beauty in one's equations than to have them fit experiment." Paul Dirac: The evolution of the Physicist's Picture of Nature *Scientific American* 208 (5) (1963)]

XII. Literature

[1]https://www.studocu.com/ro/document/universitatea-politehnica-din-timisoara/special-mathematics-ariesanu/am2teoria-campurilor/96944775 [open 16.07.2025 - 23:23]

[2]https://mec.tuiasi.ro/diverse/sem_cdif_mec.pdf [open 18.07.2025 - 00:23] Aplicatii ale calculului diferential

[3] M. Rosculet – Analiza Matematica, EDP, Bucuresti, 1982

[4] J. A.C. Ambrósio, P. Nikravesh – Elasto-plastic deformations in multibody dynamics, February 1992, Nonlinear Dynamics 3(2):85-104, DOI:10.1007/BF00118987

[5] https://en.wikipedia.org/wiki/Elasticity_tensor [open 17.07.2025 - 20:17]

[6] https://en.wikipedia.org/wiki/Finite_strain_theory [open 23.07.2025 - 22:41]

[7] E. Absi La Théorie de la plasticité et l'équilibre limite en mécanique des sols, Annales de l´Institut Technique du batiment et des travaux publics, Nr. 421, Janvier 1984

[8] M. Florentina – Rezistenta Materialelor, https://mec.tuiasi.ro/diverse/FMRM2.pdf [open 09.08.2025 - 01:23]

[9] J. Tinsley Oden – A SHORT-COURSE ON NONLINEAR CONTINUUM MECHANICS CAM 397, Introduction to Mathematical Modeling Third Edition, Fall 2008

[10] R. M. Hackett - Hyperelasticity Primer, Springer, 2018

[11]M. Crasmareanu – Geometrie Diferentiala, 28 Aprilie 2025, https://www.math.uaic.ro/~mcrasm/depozit/Geo2_BOOK.pdf [open 10.07.2025 - 02:43]

[12] D. G. Ellwood – Mechanical metallurgy, New York, McGraw-Hill ,1961

[13] L. A. Zhikharev – A Sierpiński triangle geometric algorithm for generating stronger structures, 2021 J. Phys.: Conf. Ser. 1901 012066